Re

Disinfection By-products in Drinking Water
Current Issues

Disinfection By-products in Drinking Water

Current Issues

Edited by

M. Fielding
Water Research Centre, Marlow, UK

M. Farrimond
Severn Trent Water, Birmingham, UK

ROYAL SOCIETY OF CHEMISTRY

The proceedings of the International Conference on Disinfection By-products: The Way Forward held on 29 March to 1 April 1998 at Robinson College, Cambridge, UK

Special Publication No. 245

ISBN 0-85404-778-6

A catalogue record for this book is available from the British Library

Published by The Royal Society of Chemistry,
Thomas Graham House, Science Park, Milton Road,
Cambridge CB4 0WF, UK

For further information see our web site at www.rsc.org
Printed and bound by MPG Books Ltd, Bodmin, Cornwall, UK

Preface

It seems appropriate to publish the proceedings of a conference on disinfection by-products at this time because we have more or less reached the 'silver anniversary' of the start of the concern over the health risks that disinfection by-products in drinking water may pose.

In the last 25 years, disinfection by-products have had a major impact on several activities. They have had a considerable impact on drinking water standards, how we treat water and on monitoring. The disinfection by-product issue affects three groups. There is, of course, the consumer, who wants drinking water that is pleasant to drink and is 'safe'! At a reasonable cost. Then there is the regulator, who has to set rules and standards to ensure that drinking water meeting the standards will be safe. But we all know that safety is a relative term so the regulator has to decide 'how safe' and then set, hopefully, sensible, practical standards. And finally, the water supplier has to treat water to produce 'safe and pleasant' drinking water that meets the standards – at a reasonable cost.

Underpinning all of this activity is the generation of scientifically-sound information. Good data from good research. But it is impossible to research everything of interest related to disinfection by-products. The money is not available. So what has to be decided is where the focus for disinfection by-product research should be placed. What information is still needed? And how does one balance day-to-day research requirements with more speculative precautionary investigations?

Balance is a word that crops up continually when trying to understand the disinfection by-product issue. For example, one may want to try to strike a balance between controlling disinfection by-products and incurring excessive costs. And of utmost importance is striking a balance between the desire to control disinfection by-products and compromising disinfection. And also striking a balance between setting stringent standards and the continued utilisation of important treatment techniques.

These proceedings cover a stimulating discussion of the current issues in relation to the formation, control, significance and regulation of disinfection by-products. The papers cover the views of researchers, regulators and water suppliers from Europe and North America and as such help to provide a balanced view of the current challenges posed by disinfection by-products and some idea of the way forward.

Mike Farrimond
Mike Fielding

Contents

Advances in Analysis and Monitoring

Standards and Regulation

Balancing Chemical and Microbiological Risk

Control of DBPs

Concluding Remarks

Introduction

WATER IN EUROPE

Th.G.Martijn,

President of EUREAU
European Union of National Associations of Water Suppliers and Waste Water Services.
Brussels.

1 INTRODUCTION

It is a great pleasure for me, as President of the European Union of National Associations of Water Suppliers and Waste Water Services Eureau, to address you at the beginning of your conference. EUREAU's job - as defined in its statutes - is to represent the common interests of its member associations to the European Community organisations dealing with community legislation and European standards relevant to water supply and waste water. EUREAU has no power of decision, but its representative nature and its scientific, technical and managerial expertise make the organisation an appropriate body to be consulted and recognised by Community organisations. It seeks to put its members' expertise and knowledge at the disposal of the Community organisations, so that they may be taken into account in any new legislation affecting the water industry, ensuring that consumers' interests are properly considered. EUREAU's remit is twofold: to review and discuss prospective legislation and standards to give Community organisations the collective view of the water industry across Europe and to analyse existing legislation so that, at the time of revision, a sensible view can be put forward, balancing the politically desirable with the practically achievable.

I will say some words about the forthcoming revised Drinking Water Directive and the Proposal for a Framework Directive for Community Action in the field of Water Policy. But first of all some remarks concerning the world water situation, in order to put our western European problems in a more global perspective. Only a tiny fraction of the water which covers the earth is of use to humanity: 97 % is saline, filling the oceans and seas. Of the remaining 3 %, 99 % is out of reach - frozen up in icebergs and glaciers and buried deep underground. We depend on what is left to quench our thirst, wash away our wastes, water our crops and, increasingly, to power our industries. In most parts of the world this limited supply is overstrained. Industrial wastes, sewage and agricultural run off, rivers and lakes overloaded with chemicals, wastes and nutrients, threaten potential and existing drinking water supplies.
In a world population of some 5000 million people, more than a third do not have safe

drinking water and a quarter do not have sanitation. Some 50,000 deaths occur every day from waterborne diseases, which is a very sobering thought for those of us engaged in the potable water industry.

We all share the view that water is a vital and precious commodity. Adequate water supplies make an essential contribution to public health. Worldwide experience shows that communities which are deprived of proper water services become subject to water related diseases of various kinds. Even in the recent past cholera outbreaks have been reported in countries where people lack these basic facilities. Within this context the resolution from the 1977 United Nations Conference at Mar del Plata stated that:

'All people have the right to have access to drinking water in quantity and of quality equal to their basic needs'.

This conclusion led to the decision by the United Nations to establish the period from 1981 - 1990 as the International Drinking Water Supply and Sanitation Decade. The aim of the decade, to supply everyone on earth with a minimum of safe drinking water by 1990, was entirely praiseworthy but too ambitious and unrealistic. At the end of the decade it was estimated that some 700 million people have been supplied with drinking water and some 500 million people with appropriate sanitation. Unfortunately, although the number of people supplied has increased, the number without facilities has also increased due to the increase in population in the countries concerned. Population increase has been approximately equal to the progress made. Appropriate technology and engineering has been used to attempt to supply water even in the minimum essential quantity and quality, and standards have been set at much lower levels than in industrialised countries. This conference on "Disinfection by-products" addresses one of the major problems water suppliers in the developed world have to face. But in comparison with the problems faced in the developing countries it is a luxury problem.

Soon the revision of the Drinking Water Directive will be completed. It has been a long procedure to revise the existing directive EC 778. For those who are not familiar with Community legislation it is worthwhile to explain the EU legislation procedure briefly. The EU environmental policy has two main objectives:
- it has to contribute to the progressive integration of EU Member States, and
- it has to try to achieve the objectives fixed in article 130 R of the European Union Treaty, namely:
  - to protect and improve the quality of the environment
  - to contribute towards protecting human health
  - to ensure a prudent and rational utilisation of natural resources.

The European Commission is the only EU institution that can propose legislation, ensuring the objectives of the Community's founding treaties. Draft proposals are drawn up by the Commission services and subsequently adopted as formal proposals by the College of twenty Commissioners. The College of Commissionars then makes formal proposals to the Council of Ministers, which adopts the proposals or modifies them. Past experience shows that the cases where proposals from the Commission have not been adopted are rather rare. A major legal instrument to initiate Community legislation is a directive. Directives stipulate the results to be achieved, but leave the form and methods to Member States. Directives must be transposed into national legislation.

The most important directive for water suppliers is of course directive EC 778 on the quality of water intended for human consumption, which was adopted in 1980. This directive laid down for the first time a set of mandatory quality standards for drinking water throughout the Community, which Member States were obliged to meet. The directive was based on scientific knowledge of the 1970s but the approach then adopted was too inflexible and did not take account of the wide variety of situations that existed in different parts of Europe. This variety of situations became even more important with the enlargement of the Union. Therefore, all interested parties felt that there were reasons for changing the directive. In December 1993 the Council of Ministers asked the Commission to revise the Drinking Water Directive; also in 1993, the World Health Organization drafted new recommendations on the quality of drinking water. These recommendations and the EUREAU reports on the subject were used by the European Commission as a basis for redrafting the directive. In order to avoid the previous criticism on the existing directive, the European Commission was anxious to ensure a full and public debate of relevant issues in drawing up proposals for a new Directive. EUREAU was, and still is, in regular contact with the services of Directorate General XI on technical issues.

Based on this consultative process a proposal for a new directive concerning the quality of water intended for human consumption was adopted in January 1995 and submitted to the European Parliament. In general, the new Directive has been welcomed by a variety of interests, including the water sector, as representing a major improvement of the old directive. Taking into account a number of amendments adopted by the European Parliament in its December 1996 plenary session, the Commision issued in June 1997 an amended proposal for a Council Directive on the Quality of water intended for human consumption. In February 1998 the Council of Ministers send a final proposal, the so called Common Position, for a second reading to the European Parliament. After that second reading the Council can reach a final agreement. It is foreseen that this year the revision of the Drinking Water Directive will be completed.

2 THE CONTENTS OF THE PROPOSED DRINKING WATER DIRECTIVE

Reduced number of parameters

In the light of the subsidiarity principle which was introduced in the new Treaty of the European Union signed in Maastricht, and which means that decision-making should be decentralised as far as possible, the total number of water quality parameters has been reduced from 67 in the existing Directive to 48 in the proposal. This includes 13 new parameters which have been added in the light of progress in scientific understanding. In accordance with the subsidiarity principle, only those parameters considered essential at the level of the Union were included, leaving Member States free to add secondary parameters if they see fit. This reduction of parameters will not, however, lower the level of health protection.

Microbiological parameters

These are key microbiological parameters for which no derogation is permitted. This parameter group includes E-coli and Faecal streptococci.

Chemical parameters

This group currently contains 26 parameters. Key issues are lead and pesticides. The main issues debated were the advisability of setting a limit for total pesticides as well as for individual ones and the cost of replacing lead piping. Concerning pesticides, the existing EU standard of 0.1 ug/l is retained as a common precautionary value, even though this is not based on toxicological advice. The value of 0.5 ug/l total pesticides has been reintroduced in the Common Position at the request of the European Parliament. Concerning lead, the proposed standard has been reduced from 50 ug/l to 10 ug/l, based on WHO guidelines. According to a Commission study it might cost 34 billion ECU to replace lead piping in the European Union. The standard of 10 ug/l should be met 15 years from the date of entry into force of the directive and the standard of 25 ug/l should be met after 5 years. Possible derogations give Member States some flexibility in the time schedule.

Indicator parameters

These comprise chemical and microbiological parameters which are not necessarily of direct health significance but may indicate water quality problems, including aesthetic quality.

Monitoring

The monitoring requirements of the new Drinking Water Directive seek to strike a balance to achieve maximum value from the expensive investment in sampling and analytical facilities. It is recognized, for example, that not every parameter needs to be frequently monitored, particularly when a substance is unlikely to be present or to exceed the standard. The directive also lays down criteria for sampling and analytical accuracy. For most parameters the directive specifies target levels of accuracy, precision and limit of detection.

Technological consequences

As said before, one of the major problems water suppliers will have to face is the formation of disinfection by-products. It is indeed paradoxical that, by treating the water, new and undesirable substances are formed. However, disinfection before distribution is in some countries extremely important to avoid any microbiological risk. Strict values for chemical parameters should not have the indirect consequence of increasing the risk for microbiological contamination and bringing human health in danger.

3 THE WAY FORWARD AND OTHER IMPORTANT CONSIDERATIONS

This conference brings together water supply experts, regulators, research scientists and environmentalists to consider the latest developments in research, legislation and management in relation to the difficult, and often conflicting, disinfection by-product issues. I sincerely hope that the outcome will be a clearer view of the way forward.

Enough about drinking water and public health. Water is also a pre-condition for a sustained economic development. Industrial, commercial and agricultural sectors cannot flourish without the availability of this vital resource. There exist impressive irrigation infrastructures as a requirement to sustain agriculture throughout the year. Also industries put their specific requirements on quality and availability of water.

I also wish to emphasize the role of water for the natural environment. We have seen too many exemples in the past of a declining natural diversity. Environmental harm was caused for instance by deterioration of the surface water quality or by abstractions of groundwater beyond the safe yield and consequently the lowering of water tables. People however tend to learn from their experiences and mistakes. That is why fortunately these days the implementation of large water projects is made dependent, more or less as a matter of routine, of the outcome of environmental impact assessment studies. In the past decades also   the need to consider sanitation, which includes the collection and treatment of waste water, has increased. Without the proper treatment of domestic and industrial wastewater the quality of receiving water bodies will inevitably deteriorate. This consequently complicates its use as a source for the production of drinking water.

Appreciating the vital role of water for the well-being and the economic development of a society, we observe that water is subject to conflicting interests. Many sectors put their demands on scarce available water resources. The intermediary role of government conse-quently becomes crucial to arrive at a fair distribution of this resource among the different user groups. Effective water management, embodied in legislation, policies and planning, becomes especially crucial in environments which are characterized by high population densities and economic development. Water management is a subject which during the past decades has strongly gained importance. It requires the co-operation between policy makers and research bodies to arrive at appropriate solutions which satisfy the requirements of all stakeholders. Effective water management cannot do without detailed studies to assess the capacity of available water resources, to forecast future water demands, and to develop and rank alternative technical and policy scenarios for meeting water demands. It also requires effective co-ordination between central and local governments based on procedures, rules and regulations, for instance for physical and water resources planning. A water management system finally requires inspectorates to supervise the enforcement of the legislation which has been put into place. Experience shows that without such system, water tables in rural areas may fall markedly each year, because groundwater is over-abstracted for use in irrigation. Without proper management structures it will also be increasingly difficult to provide adequate water and sanitation services to the millions in this world, who are accommodated in the ever growing number of cities and suburbs.

For long time the public has considered the availability of water as a gift from the gods. In parts of the world there is consequently still a reluctance of the population to pay for service delivery. There are other examples which show that tariffs are kept low because of political considerations. It is now generally recognized that the cost of providing water services, be it water supply or the collection and treatment of wastewater, must be recovered from the consumers. Unless this is done, operation and maintenance of the water infrastructure will suffer, as will the level and quality of the service. Water is thus to be considered as an economic good, for the use of which an adequate price must be paid. The countries under transition in Eastern Europe provide the example of water service provision which was heavily subsidized by the central government. The delivery of water at hardly any cost to the consumers has however resulted in irresponsible water use and wastage. That is why water companies should be allowed a financially independent position and be required to recover the exploitation and investment cost directly from the consumers.

This, and more, was in the mind of the European Commission when it recently produced a proposal for a new Framework Directive for Community Action in the field of Water Policy (hereafter referred to as the Water Policy Framework Directive). The proposed Water Policy Framework Directive is the outcome of a long discussion with active participation by EUREAU. The proposed new directive provides a major step forward and contains a number of key features warmly welcomed by EUREAU. These include:

- The establishment of a framework for the protection of water with common objectives, principles and basic measures.

- The holistic management of water resources, both surface and groundwater and qualitative and quantitative aspects.

- The reference to both the intended use of water resources and the intrinsic protection of aquatic ecosystems.

- Reference to the principle of subsidiarity.

- Looking at all sources of pollution: point, diffuse and accidental.

- The inclusion of transparancy and public availability of data.

- The review of older and out of date directives.

Although not directly in relation with the subject of this conference, I thought it relevant to elaborate on this important development in European Water Policy. It will be a while before the proposed Water Policy Framework Directive becomes Member State legislation.

EUREAU is in constant dialogue with the Commission organisations to explain its views and hopes that steps forward will be made to a sustainable environment and sustainable drinking water supply in Europe.

DISINFECTION BY-PRODUCTS - A VIEW FROM NORTH AMERICA

Richard J. Karlin

Deputy Executive Director
AWWA Research Foundation
6666 West Quincy Avenue
Denver, CO 80235

1 INTRODUCTION

Ever since Dr. John Snow established in the mid-1800s that water could be a mode of transmission for such deadly diseases as cholera, water supply professionals have been devising methods for interrupting this transmission. As early as 1893, Brewster, New York, US, applied chlorine to sewage effluent as a means for protecting the New York City water supply. In 1897, in Maidstone, Kent, in the UK chlorine was used for the first time to disinfect drinking water when it was applied to a water main associated with a typhoid outbreak. The outbreak was reportedly associated with pollution of the water supply by hop pickers. It seems unlikely to me that anyone serving such a noble purpose as picking hops could be associated with any adverse result, but that is another story. In 1908, Chicago used chlorination of the drinking water to stop a typhoid epidemic, and the die was cast. Chlorination of drinking water to prevent disease became one of the most widely practiced public health measures in the developed world and may be the most effective health measure undertaken in the twentieth century. In fact, it is credited with reducing US cholera incidence by 90 percent, typhoid by 80 percent and amoebic dysentery by 50 percent. No mean feat, the premature deaths avoided by chlorination of drinking water easily outweigh what was later determined to be the potential health risk posed by the chemicals produced in association with the chlorination process.

For every silver lining there is a dark cloud. Although it took some time for the issue to emerge, the discovery in the mid-1970s of by-products of the chlorination process, such as chloroform and other trihalomethanes, created quite a stir. Several of these chemicals have the unfortunate attribute of causing unpleasant health outcomes in rodents, especially in rodents bred specifically to be sensitive the these outcomes. Of especially high concern was the fact that at high dose levels various organ cancers, especially liver cancers, can occur in these rodents. If you make the assumption that "mice are little men," and/or that "men are big mice" the implications are significant. The fact that most of the cancer sites of concern in our rodent friends are very rare in humans coupled with the fact that the later epidemiological studies for the most part identified a different endpoint of concern than the rodent studies hardly adds clarity to an already unclear picture.

2 REGULATION OF DISINFECTION BY-PRODUCTS

It was concern over the potential for human carcinogenicity that was the initial driving force for the promulgation of regulations for the control (or at least the limitation) of trihalomethanes in public drinking water. The details of these regulations are best left for definition in another paper being presented at this conference. Suffice to say the honeymoon with chlorine disinfection of drinking water was, if not over, at least in need of marriage counseling. This threat to the future of chlorination was no small matter. 70 percent of US systems serving >10,000 people and 95 percent of systems serving <10,000 people rely on chlorine for primary disinfection.

The pivotal issue is that, somehow or another, water suppliers, for the first time, were required to provide water that not only met microbiological standards but disinfection by-products standards as well. There is a somewhat widespread assumption in the US that if a little of something is good, a lot must be better. For example, chemical fertilizer and, unfortunately, chlorine were sometimes applied at levels well in excess of those required to accomplish their intended purpose. At the very least, the regulation of disinfection by-products forced water purveyors to examine their chlorination practices and optimize the process.

An unintended by-product of the fuss over by-products was that the water supply community "forgot" why disinfection was undertaken in the first place. Although the details of various versions of the story abound, it is fairly widely accepted that the 1991 cholera epidemic in Peru was the result of changing disinfection practice in order to minimize the risk posed by by-products at the expense of microbiological safety. In this case, the decision was made to trade the real risk of potentially fatal microbial infection for the still somewhat hypothetical threat of disinfection by-products. In retrospect, the decision hardly seems a wise one. Be that as it may, the issue of balancing the risks posed by the microbes in water and the risk posed by the process which controls them has been the principal driver for the US water supply community for several years. It can be expected remain an overriding issue so far into the next century.

If the microbiology versus by-product issue is not difficult enough, the control of various types of by-products (haloaceticacids vs. trihalomethanes(THMs)) can be just as perplexing. For example, high pH favors THMs, low pH, haloacetic acids (or haloacetates). Simultaneously maintaining high and low pH is a challenge to even the most expert of chemists. Although not directly related to disinfection by-products, it was concern over lead levels that led Milwaukee to change coagulants to help control pH (and therefore lead levels in the water); the rest is history. Reports as to the actual number of individuals involved vary. But consensus is that up to 400,000 people were infected and between 45 and 100 died as a result of the *Cryptosporidiosis* outbreak which was at least circumstantially related to the coagulant change. The point is that there is no free lunch, everything is a trade-off or, as Newton said, "to every action there is an equal and opposite reaction." Murphy has written a number of laws (too lengthy to list here) that are also at work in the microbe/by-product balancing act business.

By the early 1990s a number of human epidemiological studies regarding the relationship between exposure to chlorinated water (and inferentially disinfection by-products) and bladder/rectal cancers had been completed. While the results of these human studies have an inherent appeal over the reliance on animal models, the sum total of the findings is equivocal at best. Some of these studies have indicated a potential problem with exposure to chlorinated drinking water, others have not. The net result is that no direct causal association has been established between drinking chlorinated water and an excess risk of cancer. In 1992, a so-called meta-analysis of existing studies was conducted. This analysis pooled the results of a number of other studies in an attempt to add statistical power to the results of those studies. Unfortunately, the analysis adds little to the overall understanding of the issue. That is, the combination of a number of studies each with their own individual methodological limitations does not completely eliminate the problems inherent in the individual studies.

A 1992 epidemiological study, in New Jersey, identified a possible link between THMs and adverse reproductive outcomes. Study design issues remain. The study was actually looking for an association between such outcomes and an entirely different set of chemicals making the relationship with chlorinated water difficult to interpret. A California study released in February 1998, suggests a link between THMs (in particular bromodichloromethane) at levels commonly encountered in chlorinated drinking water and spontaneous abortions. Though this study is merely suggestive, it adds to the media/public concern over the role of chlorinated drinking water in adverse health outcomes.

The general insinuative nature of the animal and epidemiology relating consumption of chlorinated drinking water and adverse health effects led to an increased interest in the substitution of ozone (or other oxidants) for chlorine as primary disinfectant. In fact, there are those who feel that US regulators were headed toward a mandate specifying the use ozone and chloramine (for residual maintenance) as best disinfection practice, essentially forcing a nationwide shift from chlorine to ozone. In the early 1990s, a Research Foundation study identified bromate as a by-product of disinfecting waters containing bromide. Japanese studies had indicted bromate as a potential carcinogen and ozone became a suspect rather that a saviour. Other alternate oxidants also have byproducts or associated trace impurities with known or suspected adverse health effects. This is hardly surprising since these chemicals share the characteristic of being strong oxidizers. Therefore, they all tend to oxidize organic (and for that matter inorganic) matter in the raw water and to result in similar types of oxidation (disinfection) by-products. Even for chlorine, which has been extensively studied, only a fraction of the chemicals produced by the oxidation of the various forms of natural organic mater in raw waters is characterized or even characterizable. Since the other oxidants have been less extensively examined, there are fewer known bad actors in the mix of the by-products which they produce. Some experts contend that a principal reason that alternate oxidants look so attractive is that their by-products are so poorly characterized. That is, we fear chlorine because we know so much about its by-products, and trust alternates because we know so little.

Recent research has indicated the brominated by-products may be more carcinogenic (and perhaps have greater reproductive impact) than their chlorinated analogs. To date, the US regulations have included Maximum Contaminant Levels (MCLs) for total trihalomethanes rather than attempting to control DBPs on chemical-by-chemical basis. The relatively higher apparent toxicity of a subgroup of these by-products brings the future of such a scheme into question. For example, there is some evidence that a major component of the total THM concentration, chloroform, is fairly benign, at least as a carcinogen. Does this mean that a THM limit, potentially driven by schemes designed to minimize chloroform, is less health protective than a limit on bromodichloromethane, for example? If that is the case, is the public health better served by a longer list of contaminants individually controlled at a lower level than by a limit on the sum of the individual chemical concentrations? What about mixtures? Are the toxic effects of various DBPs synergistic? antagonistic?   additive?   These unknowns confounded by Congressional mandate to regulate by-products put the entire drinking water community on the horns of a proverbial dilemma.

3 BALANCE OF CHEMICAL AND MICROBIOLOGICAL RISKS

Having lost ozone/chloramine as a silver bullet, and confused by other issues, the water supply community was somewhat at a loss as to how to simultaneously prevent waterborne disease and minimize the risk(s) posed by the process installed to control that risk. Concern over the need to maintain something approaching equilibrium in the battle between microbes and chemicals led to the first of its kind (at least in the US drinking water arena) negotiated regulation of microbes and disinfectants. Begun in 1992, this so - called "reg-neg" process involved a variety of stakeholders in the regulatory development process. By identifying the various and sometimes conflicting needs of diverse stakeholders, the USEPA hoped to develop a regulation more likely to meet the needs of all of these stakeholders.   Hopefully, such an approach avoids the degeneratoin of the regulatory process into a lawsuit contest. Such a contest puts the public health issues into a morass of legal posturing unlikely to serve anyone well. The net result of the "reg-neg" was a better understanding of the uncertainties and complexities of the risk-risk trade-offs and the development of a more reasoned regulatory scheme than might other wise have evolved.

The process has resulted in a set of regulations based upon three key elements:
-   a two stage DBP rule;
-   an interim enhanced surface water treatment rule and groundwater disinfection rule (aimed at microbial control); and,
-   an information collection rule.

Stig Regli of USEPA is on the programme later to detail the rules. Briefly, they provide a framework for developing and incorporating an increased knowledge base into a regulatory framework on an established schedule.   The guiding principals for regulatory development are to avoid premature shifts in disinfectant use, and to ensure that the microbiological safety of drinking water is not compromised.

4 BY-PRODUCTS RESEARCH

A key ingredient to the success of the "reg-neg" process is development of the data and scientific information necessary to formulate a sound regulation at the end of the day. Since 1986, the AWWA Research Foundation has undertaken over $60 million in research aimed at the general goal of measuring, controlling and understanding microbes and disinfection by-products. Approximately half of this effort ($30 million worth) is aimed at measurement, prevention, control and health effects of DBPs as well as the management of the processes necessary to accomplish these tasks.

In 1995, USEPA and the Research Foundation established an unprecedented partnership in the form of a Microbial/Disinfection By-Products Council. The Council is composed of representatives of the USEPA, the Foundation's Board of Trustees (utility managers) and other participants in the "reg-neg" process (the National Environmental Health Association, the Natural Resource Defense Council, and the Association of State Drinking Water Association). This body, funded more or less equally by the Foundation and EPA, is charged with the conduct of research aimed specifically at resolving unknowns related to the development of the next stage of microbial and disinfection by-product regulations. This body has underway over $5 Million in research projects related to this critical issue.

There are research questions that are of high concern in nearly every aspect of drinking water science from source control to public involvement/customer service. The following are highlights of the Foundation's research efforts to date regarding DBP s.

Water Resources

Electronic Watershed Management Reference Manual *Camp Dresser & McKee, Inc.* Helps water suppliers maximize the benefits of watershed protection while considering state, federal, and regional authority; historical experience with best management practices (BMPs); and utility-specific water quality and watershed characteristics. Includes information on existing and pending state and federal watershed protection policies, regulations, and guidance; detailed information on the watershed management programs of 84 water utilities; valuable guidance on selected BMPs; and a comprehensive state-by-state reference manual.

Evaluation of Sources of Pathogens and NOM in Watersheds *Stroud Water Research Center, Wisconsin State Laboratory of Hygiene, and South Central Connecticut Regional Water Authority (New Haven)* Will determine the distribution and densities of *Giardia* and *Cryptosporidium*, and concentrations of NOM in watersheds, and evaluate potential sources in field studies. Will develop potential source control strategies that will mitigate the concentrations of these contaminants in influent water resulting in potential treatment savings

Water Treatment

Bromate Formation and Control During Ozonation of Low Bromide Waters *Montgomery Watson, University of Colorado, University of Illinois, University of Toronto, and University of Poitiers (France)* Will evaluate the formation and control of bromate in low bromide waters under ozone dosages capable of inactivating *Cryptosporidium*. Research partner: M/DBP Council

Case Studies of Modifications of Treatment to Meet the New D/DBP Regulation *Metropolitan Water District of Southern California (Los Angeles), Malcolm Pirnie, Inc., and Apogee Research, Inc.* Will conduct case studies at 10-15 utilities that have implemented treatment modifications to comply with the proposed Disinfectants/ Disinfection By-Products (D/DBP) Rule and Enhanced Surface Water Treatment Rule. Will document the modifications made, the associated costs, and the lessons learned through implementation

Demonstration-Scale Evaluation of Engineering Aspects of the PEROXONE Advanced Oxidation Process *Metropolitan Water District of Southern California (Los Angeles) and Montgomery Watson* Will evaluate ozone and PEROXONE (ozone in combination with hydrogen peroxide) processes to confirm pilot-plant results for taste and odor control, disinfection by-product control, disinfection, and turbidity removal. Will also determine mass transfer efficiencies of the processes and evaluate various process considerations and equipment alternatives

Destruction of Disinfection By-Product Precursors Using Photoassisted Heterogeneous Catalytic Oxidation *Michigan Technological University* Investigates two aqueous oxidation technologies, advanced photocatalytic oxidation and phase transfer catalysis, for the removal of disinfection by-product (DBP) precursors. Shows that while phase-transfer catalysis is effective for the removal of specific organic compounds, photocatalytic oxidation is more effective for DBP precursor removal. Also estimates the cost and determines the rate and degree of destruction of DBP precursors. Published in 1993

Factors Affecting DBP Formation During Chloramination *University of Houston* Will identify water chemistry and mixing conditions that promote the formation of DBPs during chloramination to provide insight into the expected behavior of full-scale plants based on pilot testing. Will also study DBP formation in water sources from a wide variety of geographic locations, and develop analytical techniques for identifying currently unknown DBPs. Research partner: USEPA

Formation and Control of Brominated Ozone By-products *University of Colorado and University of Illinois at Urbana-Champaign* Studies the impacts of water quality and treatment variables on the formation of brominated inorganic and organic disinfection by-products (DBPs) by ozone and hydrogen peroxide with ozone. Develops an analytical technique for determining total organic bromide. Also performs pilot-scale testing to define DBP profiles across a process train. Published in 1997

Impacts of Ozonation on the Formation of Chlorination and Chloramination By-Products
University of North Carolina at Chapel Hill Will evaluate the formation of halogenated and non-halogenated disinfection by-products (DBPs) resulting from (1) ozonation followed by chlorination/chloramination, (2) ozonation followed by coagulation and chlorination/ chloramination, and (3) ozonation of prechlorinated/prechloraminated water. Will use laboratory-scale studies to develop kinetic models for the formation of DBPs from these various approaches. Will also verify models at several utilities using pilot- and full-scale ozonation and chlorination/chloramination.

Ozone in Water Treatment: Application and Engineering *AWWA Research Foundation and Compagnie Générale des Eaux* Describes current applications of ozone technology in drinking water treatment in terms of purpose (such as disinfection by-product control, taste and odor control, etc.), design, installation, and operation. Includes case studies and economic considerations. Published in 1991.

Integrated, Multi-Objective Membrane Systems for Control of Microbials and DBP Precursors *Kiwa N.V., University of Central Florida, Boyle Engineering Corporation and American Water Works Service Company* Will optimize sequences of different membrane types microfiltration (MF), ultrafiltration (UF), nanofiltration (NF), and reverse osmosis

(RO) that can function as a synergistic system for removing microbiological contaminants and DBP precursors. Will address the following important issues: use of staged membranes for pretreatment, minimization of chemical pretreatment, multiple treatment objectives, process sustainability, fouling minimization, reliability, and operational considerations. Will include development of a protocol for multiple membrane applications for surface water sources. Research partner: USEPA. *To be completed in 2000.*

Modifications to the Slow Sand Filtration Process for Improved Removals of Trihalomethane Precursors *University of New Hampshire* Establishes, by pilot-plant testing of media characteristics and pretreatment schemes and full-scale sampling, the optimal capabilities of slow sand filtration to remove THM precursors in small systems. Also characterizes removed organic matter and microbial populations, determines the process's cost-effectiveness for varying raw water conditions and design parameters, and compares THM-removal performance of slow sand filtration with that of the rapid-rate process. Published in 1989.

Natural Organic Matter (NOM) Rejection by, and Fouling of, Nanofiltration and Ultrafiltration Membranes: Bench-Scale and Pilot-Scale Evaluations *University of Colorado, University of Illinois, National Institute of Standards and Technology* Will assess the mechanisms of NOM-membrane surface interactions. Will also define the applicability of bench tests in simulating NOM rejection and fouling at pilot and larger scales. *To be completed in 1999.*

NOM Adsorption Onto Iron-Oxide-Coated Sand *University of Washington* Evaluates a novel approach to remove both soluble and particulate natural organic matter from water supplies, using sand grains coated with a thin layer of iron oxide. Includes studies at both laboratory and pilot scale. Published in 1993.

Removal of DBP Precursors by Granular Activated Carbon Adsorption *Malcolm Pirnie, Inc., University of Cincinnati, and Cincinnati (Ohio) Water Works* Will evaluate the use of GAC for disinfection by-product control at six representative utilities. Will optimize performance based on pretreatment (coagulation and ozonation plus biological treatment), empty bed contact time, and blending. Will study the relationship between natural organic matter characteristics and DBP formation. Also will develop optimization guidelines and cost estimates for GAC implementation. Research partner: USEPA.

Removal of DBP Precursors by Optimized Coagulation and Precipitative Softening *University of Kansas and Virginia Polytechnic Institute and State University* Will evaluate the degree of improved DBP precursor removal by full-scale water treatment plants optimized for precursor removal rather than removal of turbidity or hardness. Will assess costs and the degree of increased precursor removal as a function of source water characteristics, season, treatment technique, and optimized treatment strategy. Research partner: USEPA

Removal of Natural Organic Matter in Biofilters *The Johns Hopkins University* Investigates the ability of a biofilter to remove natural organic matter (NOM) and assesses the likely benefits of reducing NOM by biodegradation. Also investigates preozonation and

NOM biodegradation to determine optimal ozone doses. Evaluates sand as a support medium for the microbiological population. Published in 1995.

Strategies to Control Bromide and Bromate Ion   *University of Colorado*   Will develop feasible strategies for bromate control through (1) removing bromide, (2) limiting bromate formation, and (3) removing bromate after formation. Will use bench- and pilot-scale studies to evaluate options that utilities can use to achieve bromate concentrations below a targeted level. *To be completed in 1998.*

Distribution Systems

Chloramine Effects on Distribution System Materials   *University of North Carolina at Charlotte and HDR Engineering, Inc.*   Studies the corrosion potential of combined and free chlorine on a wide range of distribution system materials. Includes results of tests on metallic and elastomeric materials commonly found in distribution systems. Published in 1993.

Chloramine Decomposition Kinetics and Degradation Products in Distribution System and Model Waters   *University of Iowa*   Will characterize the influence of water quality on chloramine decomposition rates. Will evaluate and compare chloramine decomposition rates in distribution systems with rates in model systems. Will characterize the nature of the decomposition products and quantify selected DBPs. Will also develop models and fundamental relationships for predicting chloramine decomposition rates.

Formation, Occurrence, Stability, and Dominance of Haloacetic Acids and Trihalomethanes in Treated Drinking Water   *University of North Carolina at Chapel Hill, Metropolitan Water District of Southern California (Los Angeles), Montgomery Watson, and American Water Works Service Company*   Will investigate the occurrence of haloacetic acids (HAAs) and trihalomethanes (THMs) in treated drinking water and determine conditions that contribute to the dominance of one group of disinfection by-products over the other in distribution systems. Will conduct bench and field studies in several distribution system water sources to investigate factors that contribute to the formation of THM and HAA. Research partner: M/DBP Council. *To be completed in 1999.*

Nitrification Occurrence and Control in Chloraminated Water Systems   *Economic and Engineering Services, Inc.*   Identifies the extent and causes of nitrification in utilities currently using chloramines as a disinfectant. Also identifies waters prone to nitrification problems, and the operational, chemical, and microbiological parameters that lead to an occurrence. Determines the impacts of chloramine chemistry, temperature, pH, and detention time on nitrification and develops a protocol to predict a nitrification occurrence. Published in 1995.

Optimizing Chloramine Treatment   *Economic and Engineering Services, Inc.*   Provides a chloramination optimization manual for utility use. Reviews the regulatory setting, provides a literature review, presents case studies, and describes the resource requirements of chloramination optimization strategy. Published in 1993.

Monitoring and Analysis

Characterization of Natural Organic Matter and Its Relationship to Treatability Optimizing Chloramine Treatment *Economic and Engineering Services, Inc.* Provides a chloramination optimization manual for utility use. Reviews the regulatory setting, provides a literature review, presents case studies, and describes the resource requirements of chloramination optimization strategy. Published in 1993.

Characterization of the Polar Fraction of NOM with Respect to DBP Formation *Metropolitan Water District of Southern California (Los Angeles), University of Colorado, Lyonnaise des Eaux, and University of Poitiers (France)*
Will develop methods to isolate and characterize the polar (nonhumic) fraction of natural organic matter (NOM) from different source waters and points of treatment. Will also study the reaction of the polar fraction of NOM with disinfectants to evaluate the type and yield of disinfection by-products produced.

Development of an Improved, Direct Method for the Determination of Haloacetic Acids *Midwest Research Institute* Will develop a direct method for the determination of all nine haloacetic acids using a preconcentration technique and ion chromatography. Will compare this method with current USEPA methods.

Assessment of TOC Analytical Accuracy with Reference to the Disinfectants/ Disinfection By-product Rule *Academy of Natural Sciences of Philadelphia, U.S. Geological Survey, and Philadelphia Suburban Water Company (Bryn Mawr, Pa.)* Will quantify the analytical accuracy of the total organic carbon analytical methods, including evaluation of a particulate organic carbon standard and the ability of both analytical methods to accurately measure particulate organic carbon. ***To be completed in 1999.***

Improved Methods for Isolation and Characterization of NOM *University of Washington, University of Poitiers (France), U.S. Geological Survey, HDR Engineering, Inc., SAUR, and Dynamco, Inc.* Will develop and demonstrate techniques for isolation and characterization of natural organic matter. Will emphasize new, simple approaches that can be carried out in a utility laboratory. Will evaluate techniques at 18 utilities.

Survey of Bromide in Drinking Water and Impacts on DBP Formation *University of Colorado* Characterizes bromide levels at 100 water utilities throughout the United States selected to reflect a diversity in water sources. Evaluates seasonal variations at 50 utilities. Also determines ozone- based formation potentials on all samples to give an indication of potential bromate formation. Published in 1995.

Management and Administration

Balancing Multiple Water Quality Objectives: A Technical and Managerial Approach *Camp, Dresser & McKee, Inc., Santa Clara Valley Water District (San Jose, Calif.), and San Francisco (Calif.) Water Dept.*
Will address the challenge of balancing multiple, and often conflicting, water quality objectives, such as disinfection and disinfection by-product control, taste and odor control, biological regrowth control, filtration, and corrosion control. Will also involve

simultaneous optimization of the entire treatment process train at two case study utilities. Will extend the case studies to provide a framework and guidance document for use by other utilities. *To be published in 1998.*

Health Effects

Health Effects of Disinfectants and Disinfection By-Products *Washington State University* Reviews existing literature for information on the by-products of chlorine, chloramines, ozone, chlorine dioxide, and other disinfectants. Examines the occurrence and toxicity of disinfection by-products. Also identifies areas in which more research is needed for a better understanding of disinfection by-products and of the ramifications of using various disinfectants. Published in 1991.

Update of Review of Health Effects Data on Disinfectants/Disinfection By-products *Mobull Consulting, Metropolitan Water District of Southern California (Los Angeles), and Camp Dresser & McKee* Will provide a critical review of the latest data on the occurrence and toxicity of drinking water disinfectants and disinfection by-products. Research partner: M/DBP Council. *To be completed in 1998.*

Carcinogenic Mechanisms in Rat and Mouse Hepatocytes *University of Maryland at Baltimore* Will examine whether biological systems responsible for the production of chemical free radicals in the presence of certain chemicals commonly present in drinking water (carbon tetrachloride, chloroform, trichloroacetic acid) are responsible for the unique sensitivity of the B6C3F1 mouse to the induction of liver tumors. *To be published in 1998.*

Development of Methods for Predicting THM and HAA Concentrations in Exposure Assessment Studies *Colorado State University* Will develop predictive tools for exposure assessment to trihalomethanes (THMs) and haloacetic acids (HAAs) for linkage to productive and cancer epidemiological studies. Will use representative groups of waters spanning a typical range of natural organic matter and bromide. Research partner: M/DBP Council. *To be completed in 1999.*

Direct Comparative Genotoxicity Assessment of Disinfection By-products in Drinking Water Generated From Different Disinfection By-products *University of Illinois at Urbana-Champaign* Will determine toxicity of individual DBPs, mixtures of DBPs, and concentrates of disinfection water in bacterial and mammalian cells. Will examine chlorine, ozone, and the sequence of ozone plus chlorine in addition to chloramination and chlorine dioxide treatment. *To be completed in 2000.*

Effect of Dichloroacetic Acid and Trichloroacetic Acid on Cell Proliferation in B6C3F1 Mice *Environmental Health Research and Testing Inc.* Examines the role of increased cell proliferation associated with tissue damage (cell death) in the induction of tumors in rodents treated with dichloroacetic and trichloroacetic acids. Published in Dec. 1995.

Identification and Quantitation of DNA Adducts Derived from Disinfection By-products *University of North Carolina at Chapel Hill* Will assess and develop biomarkers that could be used in human studies to quantify internal dose and predict potential health effects from exposure to specific DBPs or families of DBPs. Research partner: M/DBP Council. *To be completed in 2001.*

Relationship of Dichloroacetate (DCA)- and Trichloroacetate (TCA)-Induced Hepatic Tumors With the Induction of Peroxisomes *Washington State University* Will examine whether rodent liver tumors induced by DCA and TCA are produced by a mechanism inappropriate for extrapolation to humans or by a threshold controlled mechanism. *To be published in 1998.*

DBP Formation and Occurrence in Treatment and Distribution

DISINFECTION BY-PRODUCT FORMATION: DUTCH APPROACH OF CONTROL STRATEGIES

J.C. Kruithof

Kiwa N.V. Research and Consultancy
PO Box 1072
3430 BB NIEUWEGEIN
The Netherlands

1 INTRODUCTION

In the Netherlands one third of drinking water for public water supply is prepared from surface water. The major surface water sources are the river Rhine and Meuse.

At the end of the sixties, water treatment strategies seemed to be complete with an emphasis on the microbiological integrity of drinking water. For surface water treatment this target was ensured by chlorination steps for a large extent.

However, exactly 25 years ago, the discovery of chlorination by-products in drinking water by Rook radically affected the almost unlimited confidence in chlorine [1]. Herewith the relative serenity surrounding the philosophy of surface water treatment was disturbed, which until now has not returned.

Many strategies were pursued to restrict disinfection by-product formation. In this paper the Dutch strategy to deal with chlorination and ozonation by-products will be outlined.

2 CHLORINATION BY-PRODUCTS

Shortly after Rook's discovery a coordinated analytical investigation was performed regarding the presence of trihalomethanes (THMs) in the drinking water of all treatment plants using chlorine [2]. The maximum THM-contents in Dutch drinking water in the year 1974 are summarized in Table 1.

Table 1   *Maximum THM-content in Dutch drinking water, 1974*

Source	THM μg/l
Ground water	3.2
Bank filtered water	0.6
Dune filtrate	32
Reservoir water	242

Highest THM-concentrations were found in drinking water prepared from reservoir water with multiple chlorination steps.

Immediately treatment strategies were developed to restrict the THM-content. In the period 1974-1977 priority was given to THM-removal after chlorination.

2.1   Strategy 1974-1977: THM-removal

In the period 1974-1977 surface water treatment strategies were focussed on:
- maintainance of the total chlorination regime;
- THM-removal after formation i.e. by air stripping or GAC-filtration.

Rather soon air stripping proved to be economically unfeasible.
Some results for the GAC-research are presented in Figure 1.

Breakthrough started after 2 weeks (1600 bed volumes) while after 10 weeks only an appreciable breakthrough occurred. This moderate adsorption gave rise to elution phenomena when the influent concentration declined [3].

Thus, GAC-filtration is not a suitable treatment technique for a complete removal over long periods of time. However, when partial removal is sufficient, GAC-filtration may be appropriate.
Nevertheless new treatment strategies were pursued.

2.2 Strategy 1977-1983: restriction of prechlorination and precursor removal

In the period 1977-1983 surface water treatment strategies were focussed on:
- restriction of THM-formation by lowering pre (transport, breakpoint) chlorination use;
- precursor removal prior to post chlorination;
- maintainance post chlorination.

For all major surface water sources THM-concentration and residual chlorine were measured as a function of the chlorine dose (Figure 2).

Figure 1   *CHCl₃-levels before and after carbon filtration of pretreated Biesbosch water*

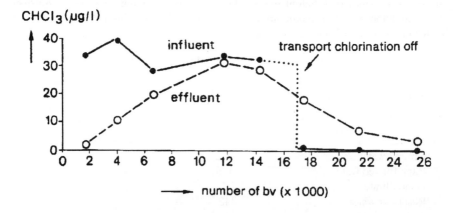

Figure 2 *THM-concentration and residual chlrorine levels after chlorination of Biesbosch water*

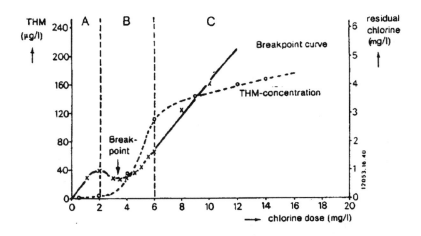

It is apparent that relatively few THM s are formed before the top of the breakpoint curve, where significant amounts of free chlorine are detectable for the first time. Immediately after the breakpoint marked THM-formation occurs, at higher chlorine dosages THM s increase only slightly. Thus three regions can be differentiated: region A (0-2 mg/l chlorine) with scarcely any THM-formation, region B (2-6 mg/l chlorine) with marked THM-formation and region C (> 6 mg/l chlorine) where THM-formation is high [4].
The dates for the effect of the restriction of the chlorine dose are summarized in Table 2.

Table 2 *Decrease of THM-formation by restriction of the chlorine dose*

	1974-1977		1978-1983	
type of chlorination	Cl_2 *mg/l*	*THM µg/l*	Cl_2 *mg/l*	*THM µg/l*
transport	4-7	48-79	1-4	7
breakpoint	5.3	51	4.1	23

For the transport chlorination the chlorine dose shifted from region B to A thereby decreasing the THM-formation significantly. Due to its purposes the dose for breakpoint chlorination is still in region B but is restricted to the minimum dose needed for disinfection purposes [4].

In this period precursor removal became a very important issue. Enhanced coagulation and GAC-filtration were investigated for this purpose. Precursor removal by GAC-filtration proved to be a very complex subject. The entire phenomenon is summarized in Figure 3.

Figure 3   *THMFP, THM-consumer and CHBr₃-consumer as a function of the DOC-content of GAC-filtrate from pretreated Biesbosch water*

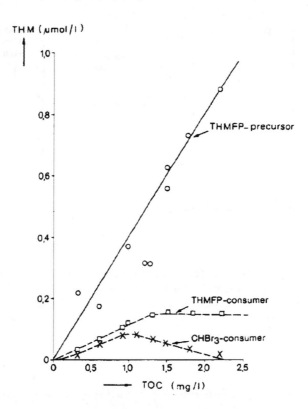

THMFP increses as a function of DOC. THM-consumer (determined at a in practice applied chlorine dose of 0.5 mg/l) increases to a certain DOC-value then remains constant. The CHBr₃-consumer rises to a somewhat lower DOC-value and then decreases [5].

Partial THMFP-reduction did not lead to a decrease in THM s under chlorination conditions customary in the Netherlands. In addition a shift took place towards the formation of more highly brominated THM s.

As a result precursor removal was given a lower priority.

In this period, in addition to THM-formation, additional side effects of chlorination were found, such as:
- a high mutagenic response in the Ames test;
- high contents of non-volatile organohalogens measured as carbon adsorbable organohalogens: AOX;
- individual chlorination by-products such as haloacetic acids, haloacetonitriles, haloketones, MX, chloropicrin, chloralhydrate, MX, etc, etc.

In view of the increasing problems with chlorination by-products Dutch water supply companies started to move away from conventional treatment with multiple chlorination steps and focussed on the suitability of ozonation [6].

3 OZONATION BY-PRODUCTS

At the end of the seventies the Dutch surface water supply companies developed the following strategy:
- replacement of chlorine by an agent with at least the same disinfecting capacity and formation of less harmful by-products;
- implementation of barriers against organic micropollutants i.e. taste and odour, halogenated solvents, pesticides, etc.;
- production of biologically stable water, thereby avoiding the necessity of post chlorination.

Treatment strategies were developed based on the implementation of ozonation followed by GAC-filtration with a strong emphasis on the microbial integrity and biological stability of the water.

3.1   Strategy 1978-1988: preparation biologically stable water

In the period 1978-1988 surface water treatment strategies were focussed on:
- increasing the disinfection capacity by ozonation;
- restriction of the mutagenic response in the Ames test. Identification of individual toxic compounds was not pursued, although the formation of bromate was established but ignored!!;
- initially characterization of biodegradable oxidation by-products i.e. aldehydes: formaldehyde, acetaldehyde, glyoxal; aldehydic acids: glyoxylic acid and carboxylic acids: formic acid, oxalic acid, etc.;
- gradually a shift from the characterization of individual biodegradable compounds to the determination of the total amount of biodegradable material expressed by the content of assimilable organic carbon (AOC);
- preparation of biologically stable (low AOC) water by granular activated carbon filtration.

For many surface water treatment plants the presence of AOC is measured in all stages of the treatment. The trend in AOC for the Kralingen treatment plant is presented in Figure 4.

After storage the AOC-content of the raw water was 30 µg/l. Coagulation caused a decrease to 10 µg/l, while ozonation caused a, by now very well known, increase to 100 µg/l. Rapid filtration and GAC-filtration caused a decrease to 10µg/l, the AOC-content recommended for biologically stable water. However, post chlorination caused another increase to 20 µg/l, an important issue when the chlorine residual did not reach the end of the distribution mains [7].

In the late eighties it was generally accepted that advanced surface water treatment including ozonation and GAC-filtration satisfied all drinking water criteria. However in the late eighties/early nineties a series of new developments took place disturbing the confidence in the application of the biological activated carbon principle.

3.2 Strategy 1989-now: protozoa inactivation and bromate control

In the late eighties/early nineties the following new developments took place:
- the analysis of pesticides in drinking water exceeding the maximum admissable

Figure 4   *AOC(P_{17}) content in all stages of Biesbosch water treatment*

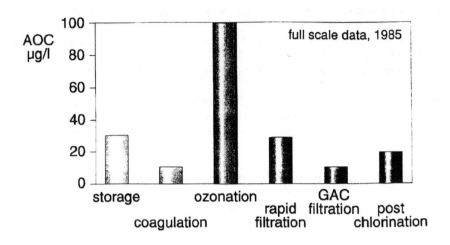

concentration in the European Union of 1 µg/l:
- the formation of bromate, a possible carcinogen to humans, by ozone and peroxone
  treatment;
- the outbreak in the USA and UK of waterborne diseases like Giardiasis and
  Cryptosporidiosis.

One of the plants considering upgrading their treatment with biological activated carbon filtration is the Berenplaat plant of the Water Supply Europoort. Much attention is paid to the balancing of disinfection requirements and bromate formation. An example for a water temperature of $15^{0}C$ is given in Figure 5.

At a water temperature of $15^{0}C$ the C.T.-requirements for 2 log Giardia elimination was 0.63 mg/l min. This C.T.-requirement went together with a bromate formation of 3 µg/l, lower than the bromate quality goal of 5 µg/l [8].

Many surface water supplies still consider biological activated carbon filtration a preferential treatment scheme for combined disinfection and organic contaminant control. They anticipate bromate formation can be restricted to values lower than
10 µg/l or even 5 µg/l.

Other surface water supplies consider to avoid chemical disinfection completely and achieve the disinfection requirements by either ground infiltration or a combination of physical treatment steps such as membrane processes.

4 PERSPECTIVE

The perspective of the Dutch surface water treatment can be illustrated by the treatment design of the Berenplaat plant of the Water Supply Europoort (Figure 6) and the Heemskerk plant of the N.V. PWN Water Supply Company of North Holland (Figure 7).

Figure 5 *Bromate formation and C.T.-value as a function of the ozone dose for pretreated Biesbosch water*

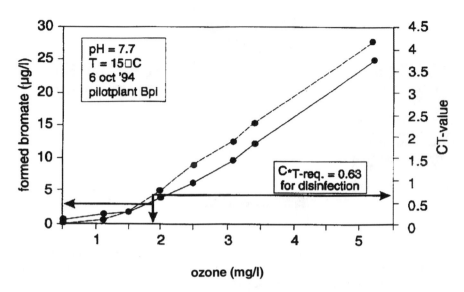

Figure 6 *Original, sand replacement and retrofit design of surface water treatment at Berenplaat*

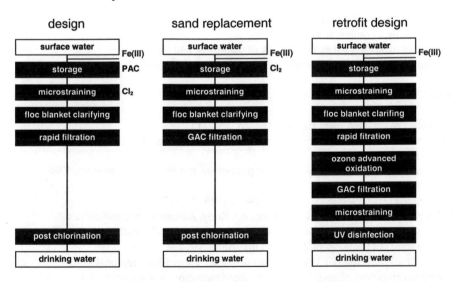

In conventional surface water treatment design, sand has already been replaced by GAC for organic contaminant control. In the retrofit design, breakpoint chlorination is replaced by ozonation for main disinfection while post chlorination is replaced by UV-disinfection. Organic contaminant control is achieved by ozonation, an optional peroxone treatment and GAC-filtration. Biological stability is achieved by GAC-filtration.

In the Heemskerk design disinfection requirements are satisfied by either ground infiltration or a combination of coagulation, rapid filtration, ultrafiltration and reverse

Figure 7   *Design of surface water treatment at Heemskerk*

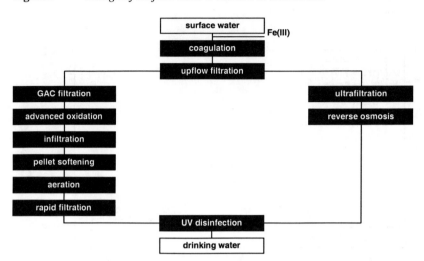

osmosis. Organic contaminant control is achieved by a combination of a non ozone based advanced oxidation (i.e. the UV/H_2O_2-proces) and GAC-filtration or reverse osmosis. Biological stability is achieved by reverse osmosis.

The Dutch surface water treatment strategy for the 21st century can be summarized as follows:

- total avoidance of chlorination;
- disinfection by a combination of processes:
  a. including ozonation balancing the risk of pathogens and DBP s;
  b. including biological and physical (i.e. membrane) processes only.

REFERENCES

1. J.J. Rook, Water Treatment Exam. 1974, **23**, 234-245.
2. J.C. Kruithof, Proceedings Trihalomethanes in water seminar, Medmenham, 1980, 211-230.
3. J.C. Kruithof, Chlorination by-products: production and control, Nieuwegein, 1986, 108-112.
4. G. Oskam, De desinfectie van water, Brussel, 1985, 415-432.
5. J.C. Kruithof, M.A. van der Gaag, D. van der Kooy, Aquatic humic substances: Influence on fate and treatment of pollutants, Denver, 1989, 663-680.
6. J.C. Kruithof, J.C. Schippers, J.C. van Dijk, Aqua, 1994, **43**, 47-57.
7. J.C. Kruithof, A.J. van der Veer, J.P. van der Hoek, Proceedings regional conference on ozone, ultraviolet light, advanced oxidation in water treatment, Amsterdam, 1996, 277-290.
8. J.C. Kruithof, E.J. Oderwald-Muller, R.T. Meijers, J. Willemsen-Zwaagstra, A.J. van der Veer, P.A.N.M. Nuhn, Proceedings regional conference on ozone, ultraviolet light, advanced oxidation in water treatment, Amsterdam, 1996, 337-351.

THE IMPACT OF NATURAL ORGANIC MATTER (NOM) ON THE FORMATION OF INORGANIC DISINFECTION BY-PRODUCTS: CHLORITE, CHLORATE, BROMATE AND IODATE

Wido Schmidt*, Ute Böhme*, Frank Sacher** and Heinz-Jürgen Brauch**
*DVGW-Technologiezentrum Wasser Karlsruhe (TZW), Außenstelle Dresden,
Scharfenberger Straße 152,
D-01139 Dresden, Germany
** DVGW-Technologiezentrum Wasser Karlsruhe (TZW), Karlsruher Straße 84, D-76139
Karlsruhe, Germany

1 INTRODUCTION

The formation of the most important disinfection by-products, the trihalomethanes (THM), is significantly influenced by the quantity as well as the type and character of the natural organic matter (NOM) dissolved in the water.[1,2] In contrast to that little is known about the role of the NOM in the formation process of the inorganic by-products *chlorite, chlorate, bromate and iodate.*[3]

In recent years several investigations have been dealing with the kinetics of the formation, especially in the case of bromate.[4,5,6,7] Nevertheless, it is difficult to explain the role of the organic matter in this process.[8,9]

The formation of the inorganic by-products can be described by the following equations:

$$ClO_2 + e^- (NOM) \Rightarrow \underline{ClO_2^-} \tag{1}$$

$$2ClO_2 + 2OH^- \Rightarrow \underline{ClO_2^-} + \underline{ClO_3^-} + H_2O \tag{2a}$$

$$3ClO^- \Rightarrow 2Cl^- + \underline{ClO_3^-} \tag{2b}$$

$$O_3 + Br^- \Rightarrow O_2 + OBr^- \tag{3a}$$

$$2O_3 + OBr^- \Rightarrow 2O_2 + \underline{BrO_3^-} \tag{3b}$$

$$BrO^-/BrOH + OH^-/O_3.... \Rightarrow ......\underline{BrO_3^-} \tag{3c}$$

$$I^- \text{ (oxidation by } O_3 \text{ or } OCl^-) \Rightarrow \underline{IO_3^-} \tag{4}$$

Chlorite is formed by the reaction of chlorine dioxide with NOM dissolved in water which plays the role of an electron donor (equation 1). Moreover, the formation of chlorite is possible by auto-decomposition of residual chlorine dioxide in alkaline medium, whereby chlorate can be determined in the same molar ratio (equation 2a).[10]

The main source of chlorate input is recognized to be the non-appropriate storage of hypochlorite stock solutions causing their decomposition to chloride and chlorate (equation 2b).[11] The formation of chlorate as a disinfection by-product after the chlorination of drinking water has not been shown so far.

Bromate formation was well studied in recent years.[4,6] It could be proved that the main cause of the occurrence of bromate during drinking water purification is the use of ozone in combination with a relatively high concentration of bromide in the raw water. In general, two paths of bromate formation could be recognized: *the oxidative mechanism and the radical one (equation 3a to 3c).*[4,6] In contrast to the ozonation, the formation of bromate is irrelevant by using chlorine or chlorine dioxide as disinfection agents.

The formation of iodate as disinfection by-product can be expected if iodide is contained in the raw water. The oxidation of iodide to iodate should be possible by chlorination and the use of chlorine dioxide too, caused by the low reduction potential $RP_{[I^-/IO_3^-]} = 0.672$ V (equation 4). The mechanism of the iodate formation is not well examined so far, but it can be assumed that bromate and iodate formation paths are similar. With the exception of the chlorite formation (equation 1), the quality and quantity of organic matter did not play a role in description the formation process of inorganic by-products up to now. Nevertheless, it can be assumed that this impact, even when it will be more indirect, cannot be neglected. Therefore, the objective of this work is to describe the influence of the NOM on the chlorite, chlorate, bromate and iodate formation.

In order to modeling this process the parameter "efficiency of by-products formation" (E_{DBP}) is defined (equation 5).

$$E_{DBP} = [C_{DBP}] / [\Delta C_{PRECUSOR}] \qquad (5)$$

The Efficiency E_{DBP} is calculated by the ratio of the level of by-product formed [C_{DBP}] to the change of the precursor concentration [$\Delta C_{PRECUSOR}$].

The precursor concentration is defined as follows:
the level of chlorine dioxide (chlorite and chlorate formation),
the level of bromide (bromate formation) and
the level of iodide (iodate formation).

2 EXPERIMENTAL

In order to reach a systematic modification of the level of organic matter dissolved in the water different amounts of activated carbon F 300 were added to the samples. The carbon fraction was separated by filtration (0.45 μm) from the water after shaking over a period of 72 h in minimum. This process is called *"selective NOM-elimination"*.

In addition to that, the samples were adjusted to the same level of NOM exactly by using the so called *"dilution water"*. This kind of water was prepared from one part of the sample by addition of an overdose (500 mg/L) of activated carbon in order to remove the organic compounds completely. After the separation of the carbon fraction all inorganic parameters were adjusted to the original sample water. This step is called *"non-selective NOM-elimination"*. For the description of the NOM-level reached by these steps the spectral absorption coefficient near 254 nm, SAC (254), and the values of the dissolved organic carbon (DOC) were used.

Table 1 gives an overview of all analytical methods used during this work.

Table 1: *Analytical methods*

Parameter	Sample-preparation	Method	Detection
chlorine dioxide	-	DPD	UV-VIS (552 nm)
chlorine	-	DPD	UV-VIS (552 nm)
ozone	-	indigo	UV-VIS (600 nm)
chlorite	-	ion chromatography	amperometric
chlorate	-	ion chromatography	conductivity with suppression
bromate	-	ion chromatography	conductivity with suppression
iodate	-	ion chromatography	conductivity with suppression
AOBr	pre-concentration with activated carbon, oxidation to bromide	ion chromatography	conductivity with suppression
DOC	filtration 0.45 μm	catalytic oxidation	IR-CO_2 (4.3 μ)

The chlorination, the dosage of chlorine dioxide and the ozone input were carried out by using stock solutions of high concentration. After defined contact periods, *in the range from 0.1 to 2 hours* and *from 24 to 48 hours* the level of by-products formed including the residual concentration of the disinfection agent were analyzed (see Table 1). The short contact time simulates the situation after the end of the water purification process (outlet waterworks) and the longer one was selected to simulate the by-products formation during the distribution of the water and at consumers tap.[12]

3 RESULTS AND DISCUSSION

Chlorite
In general, the reduction of the level of organic matter dissolved in the water causes a significant decrease of the chlorine dioxide demand connected with a lower chlorite level. For example, the diagram in Figure 1 shows the correlation of the chlorine dioxide concentration after 0.5 h contact time versus the spectral absorption coefficient SAC (254) as one characteristic attribute of the organic matter. The chlorine dioxide concentrations added were selected between 0.1 and 0.4 mg/L according to the German drinking water guide lines.[13] The level of chlorite formed was detected after exactly the same contact time. The consequent linear character of all curves, which seems to be independent on the dose of chlorine dioxide, is remarkable.

The efficiency of the chlorite formation in this water samples versus the SAC (254) is shown in Figure 2. In this diagram the chlorite formation is expressed under the consideration of the German drinking water standard: *the minimum concentration of*

Figure 1: *Chlorine dioxide demand and chlorite formation versus the SAC (254) after the reducing of NOM by different amounts of activated carbon. Contact time: 0,5 h.*

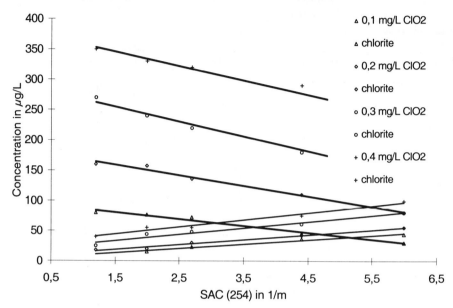

chlorine dioxide required after the end of the purification process (0.05 mg/L) and the maximum chlorine dioxide level authorized in Germany (0.4 mg/L).[13]

The efficiency of chlorite formation E_{ClO2} is increasing significantly with the reduced level of organic matter dissolved in the water if free chlorine dioxide is available (maximum dose of 0.4 mg/L chlorine dioxide). In contrast to that, the efficiency E_{ClO2} by the application of the *"minimum chlorine dioxide dose"* is constant.

The approximated linear correlation between $E_{ClO2(max)}$ versus SAC (254) and the parameter $E_{ClO2(min)}$ respectively could be reproduced for water of different origin in laboratory experiments after the activated carbon filtration. Therefore, the description of the chlorite formation in the case of the minimum as well as the maximum chlorine dioxide dose should be possible by the following equations 6a-6d:

maximum ClO$_2$-dose:

$$[ClO_2^-] = \Delta [ClO_2] \{A\ SAC\ (254) + B\} \qquad (6a)$$

A: negative constant in m;
B: positive constant;
$\Delta [ClO_2]$: chlorine dioxide demand

The negative constant A as well as the positive one B can be predicted by a laboratory experiment in which the water sample is filtered with different amounts of activated carbon as described above (compare Figure. 2: minimum and maximum dose of chlorine dioxide). The character of the correlation in Figure 2 did not change significantly with a longer contact time, although the efficiency of chlorite formation was slightly decreasing.

Figure 2: *Efficiency of chlorite formation in activated carbon filtered water using the minimum and maximum dose of chlorine dioxide versus the SAC (254). Contact time of chlorine dioxide: 48 h; maximum dose: 0.4 mg/L; minimum dose: chlorine dioxide concentration which guarantees a level of 0.05 mg/L ClO$_2$ after 0.5 h contact time.*

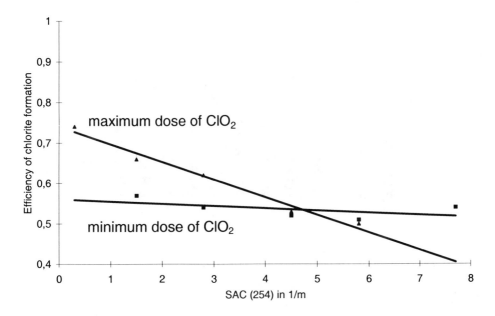

Under the conditions of a total chlorine dioxide demand, e.g. during the water distribution as usual in Germany, the correlation will be simplified. That means, the chlorite concentration expected at consumers tap can be calculated by equation 6b.

At the end of the pipe:

$\Delta [ClO_2^-] \cong [ClO_2]_0$;
$[ClO_2]_0$: dosage of chlorine dioxide
$[ClO_2^-] = [ClO_2]_0 \{A \ SAC \ (254) + B\}$ (6b)

In the case of the minimum dosage of chlorine dioxide required in Germany the chlorite level expected at consumers tap can be calculated as the product of a constant factor K and the start concentration of the disinfection agent $[ClO_2]_0$ (Figure 2:minimum dose of chlorine dioxide).

minimum ClO$_2$-dose:

$[ClO_2^-] / \Delta[ClO_2] \cong constant = K$
$[ClO_2^-] = K\{\Delta[ClO_2]\}$ (6c)

At the end of the pipe:

$\Delta [ClO_2^-] \cong [ClO_2]_0$

Figure 3: *Course of the chlorate concentration in river Elbe water near Dresden in 1996, correlation with the water flow (monthly mixed samples).*

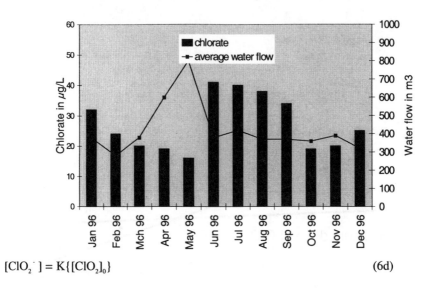

$$[ClO_2^-] = K\{[ClO_2]_0\} \qquad\qquad (6d)$$

Despite its simple character, the approximation described should facilitate a fast prediction of the average chlorite level after the disinfection at the outlet of the water works and at the consumers tap. Therefore, the model could be the base to recognize ways for the minimization of by-product formation during the distribution of drinking water.

<u>Chlorate</u>
It could be proved previously that the input of chlorate in drinking water is caused by two main paths [GORDON, 1995 and BRAUCH, SCHMIDT 1997]. These are the occurrence of chlorate in the raw water sources and the formation of chlorate in the hypochlorite and chlorine dioxide stock solutions.

The results of systematic investigations in Germany show that chlorate can regularly be found in river water and bank filtrate.[14]

The annual course of the chlorate concentration determined in river Elbe water near Dresden in 1996 is shown in Figure 3. It is remarkable that chlorate is recognized to be a permanent pollution of the river Elbe water, although its concentration level was changing significantly between 20 to 40 $\mu g/L$. A more or less indirect correlation with the water flow could be proved. Therefore, the selective input of waste waters might be the main reason of this pollution.

In order to examine the process of the chlorate formation by disinfection the different treated water samples were spiked with hypochlorite solution, absolutely free of chlorate, and fresh chlorine dioxide solution. In addition to that, some water samples were spiked with ozone prior the chlorination.

In Table 2 the concentration of the chlorate determined in the water samples is correlated with the dose of chlorine, ozone and the contact time. It is clearly seen that a change of the

Figure 4: *Chlorate concentration versus SAC (254). Dose of chlorine dioxide: 0.4 mg/L; contact time: 0.5 h*

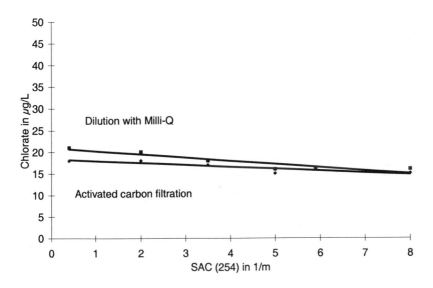

chlorate level could not be observed. Therefore, we conclude that the chlorate formation will be very low by chlorination. It is interesting too that also the ozonation in combination with a chlorination does not increase the chlorate level in the water. That means, the reaction of chlorine with organic matter or any ozonation by-products to chlorate does not take place.

Table 2: *Chlorate concentration in waters spiked with ozone and chlorine*

SAC in 1/m	DOC in mg/L	contact time of chlorine in h	ozone dose in mg/L (contact time: 1h)	chlorine dose in mg/L	chlorate in μg/L
9.0	4.6	-	-	-	12
9.0	4.6	0.5	-	1.2	14
9.0	4.6	48	-	1,2	15
5.2	3.6	48	-	1.2	14
4.6	3.2	48	-	1.2	12
4.1	3.0	48	-	1.2	18
8.7	4.8	-	-	-	12
6.9	4.8	48	0.6	1.2	13
5.7	4.9	48	1.0	1.2	13
4.4	4.9	0.1	2.3	1.2	12
4.4	4.9	0.5	2.3	1.2	14
4.4	4.9	48	2.3	1.2	13

Figure 5: *Chlorate concentration versus NOM-level (DOC) in activated carbon filtered water after different contact times. Dose of chlorine dioxide: 0.4 mg/L.*

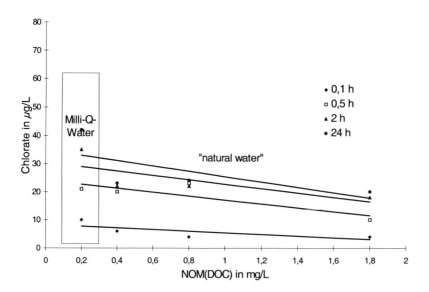

After the disinfection of the water samples with chorine dioxide the following results are remarkable concerning the chlorate formation.

The chlorate concentration measured in water with a gradual reduced level of organic matter is correlated with the SAC (254) in Figure 4. It is well seen that the curves of the activated carbon filtered and diluted water are nearly identical. Therefore, it can be concluded that the chlorate concentration is almost independent from the quality of organic matter dissolved in water. Nevertheless, a slight rise of the chlorate level with decreasing SAC (254) seems to be characteristic.

In order to examine this effect, the chlorate concentration after the dosage of chlorine dioxide was observed with increasing contact time in activated carbon filtered water. These values were compared with those determined in Milli-Q-Water (NOM $\cong$ DOC $\leq$ 0,2 mg/L). The chlorine dioxide dose added was 0,4 mg/L exactly (Figure 5).

The curves in Figure 5 confirm the slight rise of the chlorate level with decreasing DOC, according to Figure 4. In contrast to that, longer contact times cause a more significant increase of the chlorate concentration. It is remarkable that all curves are linear and approached to the corresponding chlorate level determined in Milli-Q-Water. Therefore, it can be concluded that in this case the chlorate is formed by auto-decomposition of residual chlorine dioxide mainly. The reaction of chlorine dioxide with organic matter to chlorate does not take place.

The auto-decomposition of chlorine dioxide in alkaline medium postulates a molar ratio of chlorite : chlorate = 1 (see equation 2a). It should be possible to prove this reaction in water, absolutely free of NOM. In Figure 6 the molar ratio of the chlorite and the chlorate concentration measured under alkaline conditions is shown in the case of different SAC (254)-levels and contact times. It is well seen that by removing of the NOM as the precursor substances of the chlorite formation the molar ratio is coming closer to the

Figure 6: *Molar ratio of chlorite and chlorate formed in activated carbon filtered water versus SAC (254). Dose of chlorine dioxide: 0,4 mg/L; pH-value: 8,0.*

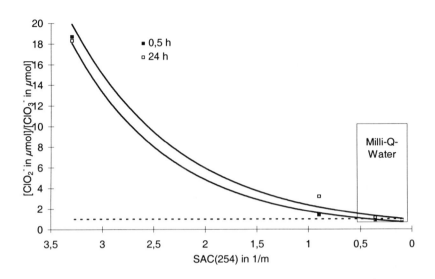

quotient 1. That means, the reaction (2a) seems to take place and should be a main source of chlorate formation. The modeling of the process of chlorate formation which depends on the pH-value should be possible and the objective of further research.

Bromate
The calculation of the bromate level formed during the water purification process by ozonation of bromide containing waters using simple models is difficult, because this process can be described by a multiparametric system only.[8] The main parameters which are to consider are the dose of the disinfection agent, the bromide level, the contact time and the pH-value.

Certainly, the level of organic matter dissolved in water seems to influence the formation of bromate in a significant way too, because it plays a key role for the ozone demand and for the formation of brominated oxidation by-products.[2,3,15,16] In general, it can be assumed that an increasing level of NOM causes a higher ozone consumption as shown in Figure 7. In this diagram the ratio of the ozone demand [ΔO_3] to the ozone dose added [O_3]$_o$ is defined as *efficiency of the ozone demand*. This quotient is correlated versus the SAC (254) as the parameter describing the level of organic matter. The sample water was prepared with different amounts of activated carbon and dilution water, respectively, as mentioned above. In both cases the efficiency of the ozone demand decreases with lower SAC (254)-values. On the other hand, it can be clearly recognized that the two curves differ from each other significantly. Especially, the quite linear character of the ozone demand in the activated carbon filtered water is worth to be mentioned.

In order to recognize the tendency of bromate formation in water with different NOM-level the samples were adjusted to 2 mg/L bromide and spiked with 2 mg/L ozone.

The correlation of the bromate concentration versus the SAC (254) is given in Figure 8. After the short contact time of 0.1 h the bromate level was not higher than 5 µg/L. A significant difference of the bromate formation in diluted and activated carbon filtered

Figure 7. *Efficiency of the ozone demand versus the SAC (254) in activated carbon filtered and diluted water. Dose of ozone: 2 mg/L; contact time: 0.5 h.*

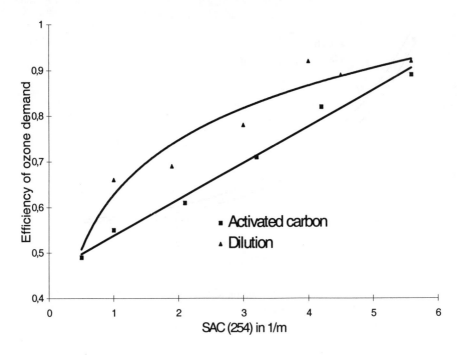

Figure 8: *Formation of bromate after ozone input in activated carbon filtered and diluted water samples versus SAC (254). Dose of ozone: 2 mg/L, bromide level: 200 µg/L.*

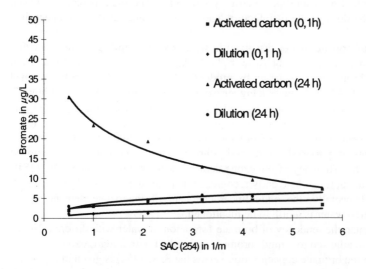

waters could not be observed (see Figure 8). On the other hand, the situation was changing completely after increasing the ozone contact time up to 24 h. Whereas the bromate level was nearly constant in the diluted water a strong rise could be measured in those water samples in which the level of NOM was reduced by activated carbon. This effect is really an amazing result and not so easy to explain. Nevertheless, it can be assumed that a competed reaction between the "three partners of reaction", *the ozone* added, *the bromide and the NOM* will take place. This might influence the bromate formation.

Figure 9: *Formation of bromo-organic byproducts (AOBr) after ozone input in activated carbon filtered and diluted water samples versus SAC (254). Dose of ozone:2 mg/l; bromide level: 200ug/l*

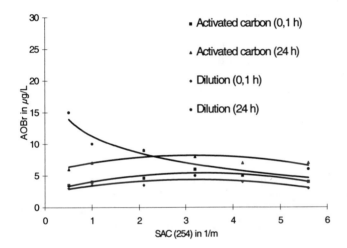

Figure 10: *Efficiency of bromate formation 24 h after the ozone input in activated carbon filtered and diluted waters.*

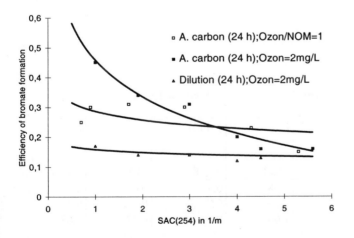

In order to demonstrate this effect, the concentration of bromo-organic compounds formed in the water samples after the ozonation, the so called adsorbable fraction of bromine containing substances (AOBr), was measured and correlated with the SAC (254) in the Figure 9.[17] The concentration of the AOBr was determined in low levels (without exception) after the short ozone contact time. After a longer contact time the AOBr-level was increasing in the diluted water samples, but in the activated carbon filtered samples not. In comparison to the tendency of the bromate formation this a completely reversed correlation and underlines the competed character between *the process of bromate formation* and *the reaction of ozone and bromide with NOM to bromo-organic compounds.* Therefore, the application of simple models is not possible.

Nevertheless, a higher bromate level can be expected by keeping up residual ozone in the water for longer time. The efficiency of the bromate formation can increase significantly after a longer ozone contact time and higher ozone concentrations in the case of reduced NOM as shown in Figure 10. The difference of the bromate concentration formed in activated carbon filtered and diluted water is remarkable.

In practice of the water purification process the activated carbon filtration is the step after the ozonation. This should guarantee a complete decomposition of residual ozone and so relatively short contact times. Therefore, it can be concluded that the impact of the character of the NOM on the bromate formation after the ozonation of bromide containing water is assessed to be relatively low and should be considered only in the case of longer ozone contact times.

Iodate

Within the group of oxygen containing halogen-ions the formation of the iodate should play a subordinated role in water treatment, because the level of the precursor "iodide" in raw waters is not higher than a few micrograms per liter normally. Certainly, it can be assumed that the rate of oxidation of iodide to iodate is high caused by the low redox-potential $RP_{I^-/IO_3^-} = 0.672V$.

The results of laboratory experiments concerning the iodate formation in iodide containing waters by ozonation and chlorination are summarized in Table 3.

Table 3: *Formation of Iodate*

(Redox-potential: $RP_{I^-/IO_3^-} = 0.672V$)
$[I^-] = 500 \ \mu g/L$;
$E_{IO_3^-} = [IO_3^-]/[\Delta I^-]$

DOC in mg/L	ozone = 1mg/L		chlorine = 1mg/L		
	$[\Delta I^-]/[I_o]$	E_{IO_3}*	$[\Delta I^-]/[I_o]$	E_{IO_3}*	E_{IO_3}*
contact time	(0.5 h)	(0.5 h)	(0.5 h)	(0.5 h)	(24 h)
2.8	≅1	1.08	0.7	0.05	0.07
1.3	≅1	1.05	0.7	0.12	0.25
0.8	≅1	1.05	0.8	0.22	0.67

* $[IO_3^-]$ calculated to $[I^-]$

The water samples, treated with different amounts of activated carbon as described above, were adjusted to 500 $\mu g/L$ iodide and spiked with ozone or chlorine.

The ozonation causes a complete oxidation of the iodide contained in water to the iodate, indicated by the ratio $[\Delta I^-]/[I_o^-]$ and the efficiency E_{IO_3}- listed in Table 3. The ratio $[\Delta I^-]/[I_o^-]$ can be defined as "iodide demand". This ratio is calculated to 1 after the ozonation and between 0.7 to 0.8 after the chlorination. The correspondence between $[\Delta I^-]/[I_o^-]$ and E_{IO_3} after the ozone-input indicates a complete transformation of iodide into iodate. Therefore, the formation of iodo-organic compounds causing the pharmaceutical taste and odour cannot be expected in this case. In contrast to that, it is well seen from the data of Table 3 that after the chlorination step the iodide demand is a little bit lower (between 0.7 and 0.8 only). The efficiency of the iodate formation is calculated between 0.05 and 0.22 only after 0.5 h contact time. Therefore, the formation of iodo-organic compounds such as iodoform is probable. Nevertheless, it is remarkable that a longer chlorine contact time and a decreasing level of organic matter causes a higher rate of iodate formation (see values of E_{IO_3} in Table 3) which is comparable to the results of bromate formation by ozonation. The transformation of iodide into iodate and iodo-organic compounds by disinfection with chlorine should be a competed reaction like the bromate formation and so the application of simple models for the description of this process is not suitable.

4 SUMMARY

The impact of the organic matter on the formation of the inorganic disinfection by-products chlorite, chlorate, bromate and iodate can be summarized by the scheme in Figure 11.

Chlorite is formed only by use of chlorine dioxide for disinfection. Its concentration

Figure 11: *General scheme of the formation of inorganic disinfection by-products*

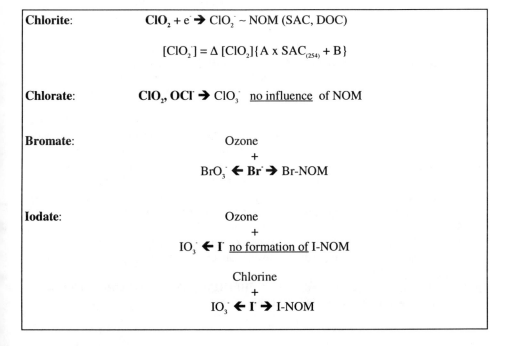

Chlorite:	$ClO_2 + e^- \rightarrow ClO_2^- \sim NOM \ (SAC, DOC)$
	$[ClO_2^-] = \Delta\,[ClO_2]\{A \times SAC_{(254)} + B\}$
Chlorate:	$ClO_2, OCl^- \rightarrow ClO_3^-$ <u>no influence</u> of NOM
Bromate:	Ozone + $BrO_3^- \leftarrow Br^- \rightarrow$ Br-NOM
Iodate:	Ozone + $IO_3^- \leftarrow I^-$ <u>no formation</u> of I-NOM
	Chlorine + $IO_3^- \leftarrow I^- \rightarrow$ I-NOM

correlates directly with the level of organic matter dissolved in the water. A simple model is developed to predict the chlorite formation at the outlet of the waterworks and at consumers tap in the case of the maximum as well as the minimum chlorine dioxide dose according to the German drinking water guide lines.

The reasons for the chlorate pollution in drinking water are its occurrence in some raw waters, the pollution of the disinfection agents, especially of hypochlorite solutions, and the auto-decomposition of residual chlorine dioxide after the disinfection. The organic matter dissolved in water does not influence the chlorate level.

The formation of bromate in bromide containing water depends on the ozone concentration which is strongly influenced by the level of organic matter. In the case of keeping up residual ozone in water with low DOC a strong increase of the bromate formation with longer contact times can be expected.

The chlorination of iodide containing water causes the formation of iodate and iodo-organic compounds. In contrast to that, a complete oxidation of iodide to iodate could be determined after the ozonation of the water.

The application of simple models in order to predict the bromate and iodate concentration in NOM-containing water of different origin will be hindered by the competed character of the XO_3^- and X-NOM-formation process.

References

1. Müller. U., Wricke, B, Baldauf, G and H. Sontheimer: THM-Formation in drinking water treatment. Vom Wasser, 80, 193 (1993).
2. Schmidt, W.; Böhme, U. and H.-J. Brauch: The formation of organo bromo compounds in treatment of waters containing bromide. Vom Wasser, 80, 29 (1993).
3. Schmidt, W.; Böhme, U. and H.-J. Brauch: Nebenprodukte der Desinfektion: Chlorit, Chlorat, Bromat und Jodat - Entstehung und Bewertung bei der Trinkwasseraufbereitung in den neuen Bundesländern. DVGW-Schriftenreihe Wasser, 86, 285 (1994).
4. von Gunten, U. and J. Hoigne: Factors controlling the formation of bromate during ozonation of bromide containing waters. J. Water SRT-Aqua., 41(5), 299 (1992).
5. von Gunten, U. and J. Hoigne: Bromate formation during ozonation of bromide containing waters. Interaction of ozone and hydroxyl radical reactions. Environ. Sci. Technol., 28, 1234 (1994).
6. von Gunten, U. and J. Hoigne: Ozonation of bromide containing waters: Bromate formation through ozone and hydroxyl radicals. In: Disinfection by-products in water treatment. R.A. Minear and G.L. Amy, CRC Press Inc., Boca Raton 187 (1996).
7. von Gunten, U. and Y. Oliveras: Advanced oxidation of bromide containing waters: Bromate formation mechanisms. Environ. Sci. Technol., 32, 63-70 (1998).
8. Siddiqui., M, Amy, G., Ozekin, K. and P. Westerhoff: Empirically and theoretically-based models for predicting brominated ozonated by-products. Ozone Sc. & Engin., 16, 157 (1994).
9. Elovitz, M. and U. von Gunten: Influence of temperature, pH and DOM source on hydroxyl radical/ozone ratios during ozonation of natural waters. 13th Ozone World Congress, Kyoto (1997).
10. Klein, E.: Analytik des Chlordioxids. CEBEDEAU, 38. International Conference of Water Disinfection. 10.-12.6.1985, Brussels.

11.Gordon, G. and L. Adam: Minimizing chlorate ion formation in drinking water when hypochlorite ion is the chlorinating agent. AWWA Research Foundation, ISBN: 0-89867-781-5 (1995).

12.DVGW-Regelwerk: Technische Regeln: Arbeitsblatt W 295: Ermittlung von Trihalogenmethanbildungs-potentialen von Trink-, Schwimmbecken- und Badebecken-wässern. (1997).

13.Verordnung über Trinkwasser und über Wasser für Lebensmittelbetriebe (Trinkwasser-verordnung- TrinkwV.) 5. 12. 1990 BGBL 1, 2313 (1990).

14.Brauch, H.-J.: et al: Systematic investigations on the significance of bromate for drinking water treatment in Germany. Report: DVGW research project, Karsruhe, Dresden, Wiesbaden 1997.

15.Schmidt, W.; Böhme, U. and H.-J. Brauch: Organo bromide compounds and their significance for drinking water treatment. Water Supply, 13(1), 101 (1995).

16.Brauch, H.-J and W. Schmidt: . et al: Minimization of chlorite and chlorate formation after disinfection with chlorine containing agents in drinking water treatment. Report: BMBF research project, Dresden, Karlsruhe 1997.

17.Schmidt, W.; Böhme, U. and H.-J. Brauch: Systematic investigations on the formation of bromate and bromoorganics in water treatment in Eastern Germany. Vom Wasser, 85, 109 (1995).

IDENTIFICATION OF DRINKING WATER DISINFECTION BY-PRODUCTS FROM OZONE, CHLORINE DIOXIDE, CHLORAMINE, AND CHLORINE

Susan D. Richardson, Tashia V. Caughran, Alfred D. Thruston, Jr., and Timothy W. Collette
U.S. Environmental Protection Agency
National Exposure Research Laboratory
Athens, GA 30605

Kathleen M. Schenck and Benjamin W. Lykins
U.S. Environmental Protection Agency
National Risk Management Research Laboratory
Cincinnati, OH

1 INTRODUCTION

Alternative disinfectants, such as ozone and chlorine dioxide, that have been widely used in Europe for decades, are now gaining popularity in the United States. Tightening regulations, along with health concerns about chlorine-containing by-products that are formed when chlorine is used to disinfect drinking water, are influencing more treatment plants to choose alternative disinfectants instead of chlorine. Of the more than 300 disinfection by-products (DBPs) that have been reported formed from chlorine treatment,[1] at least four of them--chloroform, bromodichloromethane, bromoform, and MX (3-chloro-4-(dichloromethyl)-4-oxobutenoic acid)--have been shown to cause cancer in laboratory animals. [2,3] Potential adverse health effects for many of the other reported chlorine DBPs are not known, as appropriate health effects studies have not been carried out. Alternative disinfectants typically produce much lower levels or non-detectable levels of chlorine-containing DBPs, including trihalomethanes (THMs) and haloacetic acids, allowing treatment plants to easily meet the U.S. Environmental Protection Agency's (EPA) regulations. THMs (which include chloroform, bromodichloromethane, chlorodibromomethane, and bromoform) were the first disinfection by-products regulated by the U.S. EPA, and will face tightened regulation under Stage I of the DBP Rule, with the maximum contaminant level (MCL) of total THMs being lowered from 100 μg/L to 80 μg/L.[4] Haloacetic acids (including chloro-, dichloro-, trichloro-, bromo-, and dibromo-acetic acids) will be regulated for the first time under Stage I of the DBP Rule.[4]

Although alternative disinfectants are gaining in popularity, in comparison to chlorine, there is relatively little known about the disinfection by-products that they produce. In the United States, ozone is currently used in more than 100 treatment plants, with more being designed and built; chlorine dioxide is used in more than 400 treatment plants; and chloramine is used in approximately 300 plants. Because of the significant exposure of consumers to drinking water treated with alternative disinfectants, and because there is significant uncertainty with regard to identity (and possible adverse health effects) of their DBPs, the U.S. EPA has initiated an extensive effort to determine the identity of DBPs from alternative disinfectants. Also, because alternative disinfectants are often followed by the use of a secondary disinfectant (such as chlorine or chloramine) to maintain disinfection in the distribution system, chlorine and chloramine are also being studied in combination with alternative disinfectants.

DBP studies typically target a few expected chemicals in the drinking water. Due to advanced instrumentation available in our laboratory, we are able to carry out more comprehensive DBP studies, with the ability to identify new DBPs that are not present in a mass spectral library database. In this paper, we report the successful identification of many DBPs from the alternative disinfectants and from chlorine, many of which have not been previously reported. As such, this paper represents our laboratory's effort to more comprehensively identify DBPs from different disinfectants. A more extensive review, which includes all previous DBP identifications reported in the literature, was recently published and can be found in John Wiley's *Encyclopedia of Environmental Analysis & Remediation.*[1]

2 EXPERIMENTAL

2.1 Ozone, Ozone/Chlorine, and Ozone/Chloramine Samples

Ozonated water was collected from three sources: 1) a full-scale ozone treatment plant in Valdosta, GA, 2) a pilot ozonation plant in Jefferson Parish, LA, and 3) laboratory-scale ozonations carried out on Suwannee River humic and fulvic acid-fortified distilled water (5 mg/L), with an ozone dose of 2:1 ozone:DOC. Samples were concentrated using XAD resins. Additional experimental details are similar to those in a previous study and can be found elsewhere.[5] Secondary chlorine and chloramine were applied to some ozone samples collected from the Jefferson Parish pilot plant, in amounts to achieve residuals of 2 to 3 mg/L.

Pentafluorobenzylhydroxylamine (PFBHA) derivatizations were carried out on 1 L of treated water prior to concentration, using a method by Glaze and Weinberg.[6] 2,4-Dinitrophenylhydrazine (DNPH) derivatizations were carried out using a method similar to that published by Grosjean and Grosjean.[7]

2.2 Chlorine Dioxide and Chlorine Dioxide/Chlorine Samples

Chlorine dioxide treated water was collected from a pilot drinking water treatment plant in Evansville, IN. Additional experimental details can be found elsewhere.[5] Secondary chlorine was applied to some of the chlorine dioxide treated samples, in amounts to obtain a free chlorine residual of 2 to 3 mg/L.

2.3 GC/MS and GC/IR Analyses

Low and high resolution gas chromatography-electron ionization mass spectrometry (GC/EI-MS) analyses were performed on a VG 70-SEQ high-resolution hybrid mass spectrometer, equipped with a Hewlett Packard model 5890A gas chromatograph and a DB-5 column. Low resolution experiments were carried out at 1000 resolution; high resolution experiments at 10,000. Chemical ionization mass spectrometry (CI-MS) experiments were carried out on a Finnigan TSQ 7000 triple quadrupole mass spectrometer using methane or 2% ammonia in methane gas. Gas chromatography/infrared spectroscopy (GC/IR) analyses were performed on a Hewlett Packard Model 5890 Series II GC (with

Restek Rtx-5 column) interfaced to a Hewlett Packard Model 5965B infrared detector (IRD). Additional details are found elsewhere.[5]

3 RESULTS AND DISCUSSION

3.1 Ozone

To determine the identity of ozone DBPs, we first applied a combination of GC/MS and GC/IR techniques to the XAD resin extracts of ozonated water samples. Also, methylation derivatizations were performed to aid in identification of haloacids. Overall, many DBPs were identified, several of which have never been reported previously. Table 1 lists these by-products. The majority of the ozone by-products identified contained oxygen as part of their structure, with aldehydes, ketones, carboxylic acids, aldo-acids, keto-acids, keto-aldehydes, hydroxy-ketones, and cyano-aldehydes observed. No halogenated by-products were found. Many of the compounds were not present in any spectral library (NIST or Wiley), and many of the ones that were in the libraries did not give conclusive library matches. For many of the compounds, little information was provided in the mass spectra, due to the absence of molecular ions, which provide molecular weight information. As a result, we used chemical ionization mass spectrometry (CI-MS) frequently to generate molecular ions. High resolution electron ionization mass spectrometry (EI-MS) was an indispensable tool for determining structures, as it provided the necessary empirical formula information for the molecular ion and fragments. It also helped to limit the number of possible structures for each unknown DBP. GC/IR was useful for determining the functional group. When available, standards were purchased to confirm difficult identifications and to determine a particular isomer precisely, when spectra were not conclusive. In this way, both GC retention times and spectra of the standards were matched with those of the unknowns. It should be noted that, although we attempted to identify every DBP detected in the samples, several DBPs were present at such low concentrations (less than 1 ppt) that there was not sufficient spectral information to enable their identification.

Because it is believed that many of the unidentified DBPs from ozone are located in the polar fraction, which cannot be easily extracted from water, we also applied two derivatization procedures coupled with either GC/MS or liquid chromatography/mass spectrometry (LC/MS). The first derivatization procedure involved derivatization with pentafluorobenzylhydroxylamine (PFBHA),[6] which reacts with small, polar aldehydes and ketones to form nonpolar oximes that can be easily extracted into an organic solvent, such as hexane. The derivatized ozone by-products can then be analyzed by GC/MS. Using this procedure, we found several aldehydes and ketones that had not been previously reported, and successfully identified them through the use of GC/CI-MS and high resolution GC/EI-MS. To determine specific isomers of the derivatized aldehydes and ketones, the analyses of standards was often necessary, as many isomers exhibited nearly identical mass spectra. Also, many identifications were confirmed with standards (purchased or synthesized when not available).

The second derivatization procedure we applied to identify new polar DBPs involved derivatization with 2,4-dinitrophenylhydrazine (DNPH), followed by LC/MS analysis, which had not been applied previously for identification of DBPs. Like PFBHA, DNPH also reacts with the carbonyl group of aldehydes and ketones, this time forming a

Table 1 *Ozonation DBPs Identified*

Aldehydes		Ketones	
1.	Formaldehyde [c,p]	1.	Acetone [c,p]
2.	Acetaldehyde [c,p]	2.	2-Butanone [c,p]
3.	Propanal [c,p]	3.	3-Methyl-2-butanone [c,p]
4.	Butanal [c,p]	4.	2-Pentanone [c,p]
5.	2-Butenal [c,p]	5.	3-Hexanone [c,p]
6.	Pentanal [c,p]	6.	2-Hexanone [c,p]
7.	Hexanal [c]	7.	3-Methyl cyclopentanone [c,p]
8.	2-Hexenal [c,p]	8.	Heptadecadienone
9.	Heptanal [c]	9.	6-Methyl-5-hepten-2-one
10.	Benzaldehyde [c]		
11.	Octanal [c]	**Hydroxy-ketones**	
12.	Nonanal [c]	1.	3-Hydroxy-2-butanone [c,d]
13.	Decanal	2.	6-Hydroxy-2-hexanone [c,p]
14.	2-Methyl undecanal [t]	3.	1,3-Dihydroxyacetone [c,d]
15.	Dodecanal		
16.	Tridecanal	**Di-Carbonyls**	
		1.	Glyoxal [c,p]
Cyano-aldehydes		2.	Methylglyoxal [c,p]
1.	Cyanoformaldehyde [p]	3.	Dimethylglyoxal [c,p]
		4.	Ethylglyoxal [t,p]
Carboxylic Acids		5.	2-Butenedial [t,p]
1.	2-Methyl propanoic acid	6.	5-Keto-1-hexanal [c,p]
2.	Pentanoic acid	7.	3-Cyano-2-ketopropanal [t,p]
3.	2-Methyl pentanoic acid		
4.	*tert*-Butyl maleic acid	**Keto-acids**	
5.	Benzoic acid	1.	Pyruvic acid [c,d]
6.	Hexadecanoic acid	2.	3-Keto-butanoic acid
Aldo-acids		**Other By-products**	
1.	Glyoxylic acid [c,d]	1.	Benzeneacetonitrile

[p] Identified with PFBHA-GC/MS   [t] Tentative Identification
[d] Identified with DNPH-LC/MS   [c] Confirmed Identification (with standard)

hydrazone derivative. In preliminary studies, we found that derivatization coupled with LC/MS offered significant advantages over direct analysis of the ozonated water by LC/MS for identifying new polar DBPs. The advantage that DNPH-LC/MS analyses provide over PFBHA-GC/MS analyses comes when analyzing for highly polar compounds (not amenable to GC) and for compounds that do not show clear molecular ions by GC/MS. Using this approach, we were able to successfully analyze polar aldehydes and ketones (using a single-step derivatization) that did not work well by PFBHA-GC/MS, in addition to successfully detecting the traditional aldehydes and ketones that have been previously reported as ozone DBPs. Detection limits of 0.1 ppb were obtained. Examples of polar compounds that worked well by this procedure are shown below; those compounds that were also found to be actual ozone DBPs are marked with an asterisk and were confirmed by the analysis of authentic standards.

1,3-Dihydroxyacetone*	Pyruvic acid (2-ketopropanoic acid)*
2,5-Dihydroxybenzaldehyde	Oxalacetic acid (2-keto-1,4-butanedioic acid)
3-Hydroxy-2-butanone*	Glyoxylic acid (2-aldo-ethanoic acid)*

3.2 Ozone/Chlorine and Ozone/Chloramine

When chlorine or chloramine was used as the secondary disinfectant (after treatment with ozone), a greater variety of chlorinated and brominated DBPs was found (Table 2). Many of these compounds have been identified in previous chlorination studies. It should be noted that many of the compounds were identified in a different form than the form that would actually be present in water. For example, although the haloaldehydes are listed as such, they are likely to by hydrates (as with chloral hydrate) in water. In general, there were fewer halogenated by-products formed by chloramine than with chlorine, and they appeared to be at lower levels (1/3 or lower the concentration as compared with chlorine).

Table 2 *Ozone/Chlorine and Ozone/Chloramine DBPs Identified*

	Ozone/Chlorine	*Ozone/Chloramine*
Halo-alkanes		
1. Carbon tetrachloride	X	
2. Chlorotribromomethane	X	
3. Bromopentane [t]	X	
4. 2,3-Dichlorobutane [t]	X	X
5. 1,2-Dichloro-2-methylbutane [t]	X	
6. 1-Chlorooctane [t]		X
7. 1,2-Dibromo-1-chloroethane	X	
Haloaldehydes		
1. Chloral hydrate	X	
2. 4-Chloro-3-keto-1-butanal	X	X
3. Bromodichloroacetaldehyde	X	
4. Tribromoacetaldehyde	X	
Haloacetic acids		
1. Dichloroacetic acid	X	
2. Dibromoacetic acid	X	
3. Bromochloroacetic acid	X	X
4. Dichlorobromoacetic acid	X	
5. Dibromochloroacetic acid	X	
Haloacetonitriles		
1. Dichloroacetonitrile	X	
2. Bromodichloroacetonitrile	X	
3. Bromochloroacetonitrile	X	
4. Dibromochloroacetonitrile	X	
5. Dibromoacetonitrile	X	

Table 2 *Ozone/Chlorine and Ozone/Chloramine DBPs (Continued)*

		Ozone/Chlorine	Ozone/Chloramine
Haloketones			
1.	1,1-Dichloropropanone	X	X
2.	1,1,1-Trichloropropanone	X	X
3.	1-Bromo-1,1-dichloropropanone	X	X
4.	1,1,3-Trichloropropanone	X	
5.	1,1,3,3-Tetrachloropropanone	X	X
6.	1,1,1,3-Tetrachloropropanone	X	
7.	1,1,1,3,3-Pentachloropropanone	X	X
8.	1,1-Bromochloropropanone	X	
9.	1-Chlorodimethylglyoxal	X	X
10.	Trichloromethyl ethyl ketone [t]	X	
11.	2,2,4-Trichloro-1,3-cyclopentenedione	X	
Haloalcohols			
1.	2-Chloroethanol	X	
2.	3-Chloro-2-butanol		X
3.	4,5-Dichloro-2-pentanol	X	X
4.	4,6-Dichloro-1,3-benzenediol	X	
Halonitromethanes			
1.	Bromonitromethane [c]	X	X
2.	Dibromonitromethane [c]	X	X
Other Halogenated By-products			
1.	2-Chloroethanol acetate	X	X
2.	1,2-Dichloroethanol acetate	X	X
3.	3-Chloro-2-butanol acetate	X	X
4.	Benzyl chloride	X	
5.	2,2,2-Trichloroacetamide	X	
6.	1,2,3,4,5,5-Hexachloro-1,3-cyclopentadiene	X	
7.	2-Bromobenzothiazole	X	
8.	2-Chlorobenzothiazole	X	
9.	Chloromethylbenzene	X	
Non-halogenated DBPs			
1.	Benzeneacetonitrile	X	X
2.	Heptanenitrile [t]		X
3.	5-Methyl-3-isoxazolamine		X
Plus all of the same aldehydes, ketones, carboxylic acids as with ozone alone			

[t] Tentative ID
[c] Confirmed Identification (with standard)

Table 3 *Chlorine Dioxide Disinfection By-products Identified*

Carboxylic acids		Chlorine-containing compounds	
1.	Butanoic acid	1.	1,1,3,3-Tetrachloro-2-propanone[t]
2.	Pentanoic acid	2.	(1-Chloroethyl)dimethylbenzene[t]
3.	Hexanoic acid		
4.	Heptanoic acid	Ketones	
5.	2-Ethylhexanoic acid	1.	2,3,4-Trimethylcyclopent-2-en-1-one[t]
6.	Octanoic acid	2.	2,6,6-Trimethyl-2-cyclohexene-
7.	Nonanoic acid		1,4-dione[t]
8.	Decanoic acid		
9.	Undecanoic acid	Aromatic compounds	
10.	Tridecanoic acid	1.	3-Ethyl styrene[c]
11.	Tetradecanoic acid	2.	2-Ethyl styrene[c]
12.	Hexadecanoic acid	3.	Ethylbenzaldehyde[t]
13.	2-*tert*-Butylmaleic acid[c]	4.	Naphthalene
14.	2-Ethyl-3-methylmaleic acid	5.	2-Methylnaphthalene
15.	Benzoic acid	6.	1-Methylnaphthalene
		Esters	
		1.	Hexanedioic acid, dioctyl ester

[t] Tentative Identification
[c] Confirmed Identification (with standard)

3.3 Chlorine Dioxide

Chlorine dioxide treatment, by itself, produced several DBPs that have not been reported previously.[5] Table 3 lists these by-products. Most of the DBPs contained oxygen in their structures: ketones, carboxylic acids, and maleic acids. Only two of the by-products were chlorinated, and no halomethanes were observed. The number of by-products we observed for chlorine dioxide was much lower than what is usually found for chlorination. When chlorine was used as the secondary disinfectant (after treatment with chlorine dioxide), a greater variety of chlorinated and brominated DBPs was found (Table 4). All of these by-products were also observed in the chlorinated control samples, indicating that they were formed by chlorine, and not from the combination of chlorine dioxide and chlorine. Many of these by-products have also been observed in previous chlorination studies.

3.4 Comparison of DBPs from Different Disinfectants

In comparing the different disinfectants studied, chlorine appears to produce the largest number of halogenated by-products. Chloramine produced the same types of halogenated by-products as chlorine, but they were fewer in number and lower in concentration than for chlorine. Ozone produced no halogenated by-products, and chlorine dioxide produced very few. The halogenated chlorine dioxide DBPs are probably due to the presence of a small amount of chlorine impurity that is often present in chlorine dioxide treatments. Non-halogenated DBPs are quite similar for all of the different disinfectants (e.g., carboxylic acids, aldehydes, etc.), indicating a similar mechanism of oxidation, with regard to the formation of these by-products.

Table 4 *Chlorine Dioxide/Chlorine Disinfection By-products Identified*

Haloalkanes		Other Halogenated Compounds	
1.	Bromodichloromethane	1.	1-Chloroethanol acetate
2.	Dibromochloromethane	2.	3-Bromopropyl-chloromethyl ether [t]
3.	Bromoform	3.	1,4-Dichlorobenzene
4.	Chlorotribromomethane	4.	2-Methyl-3,3-dichloropropenyl-
5.	Tetrachlorobutane [t]		dichloromethyl-ether [t]
		5.	1-Chloroethyl-dimethylbenzene [t]
Haloketones			
1.	1,1,1-Trichloro-2-propanone	Non-halogenated Compounds	
2.	1-Bromo-1,1-dichloro-2-	1.	Butanoic acid
	propanone	2.	Pentanoic acid
3.	1,1,3,3-Tetrachloro-2-propanone	3.	Hexanoic acid
4.	1,1,1,3,3-Pentachloro-2-	4.	Heptanoic acid
	propanone	5.	2-Ethylhexanoic acid
5.	2-Chlorocyclohexanone	6.	Octanoic acid
		7.	Nonanoic acid
Haloacetonitriles		8.	Decanoic acid
1.	Dibromochloroacetonitrile	9.	Undecanoic acid
2.	Dibromoacetonitrile	10.	Tridecanoic acid
		11.	Tetradecanoic acid
Haloaldehydes		12.	Hexadecanoic acid
1.	Dichlorobutanal [t]	13.	2-Ethyl-3-methylmaleic acid
		14.	Benzoic acid
		15.	Hexanedioic acid, dioctyl ester

[t] Tentative ID

References

1. S.D. Richardson, 'Drinking Water Disinfection By-products', *Encyclopedia of Environmental Analysis & Remediation*, Vol. 3, pp. 1398-1421, John Wiley & Sons, New York, 1998.
2. R.J. Bull, F.C. Kopfler, 'Health Effects of Disinfectants and Disinfection By-products', AWWA Research Foundation, Denver, CO, 1991.
3. H. Komulainen, V.-M. Kosma, S.-L. Vaittinen, T. Vartianinen, E. Kaliste-Korhonen, S. Lotjonen, R.K. Tuominen, J. Tuomisto, *J. Nat. Cancer Inst.*, 1997, **89**, 848.
4. F.W. Pontius, *J. Am. Water Works Assoc.*, 1996, **88**, 16.
5. S.D. Richardson, A.D. Thruston, Jr., T.W. Collette, K.S. Patterson, B.W. Lykins, Jr., G. Majetich, Y. Zhang, *Environ. Sci. Technol.* 1994, **28**, 592.
6. W.H. Glaze and H.S. Weinberg, 'Identification and Occurrence of Ozonation By-Products in Drinking Water', American Water Works Association Research Foundation: Denver, CO, 1993.
7. E. Grosjean and D. Grosjean, *Intern. J. Environ. Anal. Chem.* 1995, **61**, 47.

THE INFLUENCE OF UV DISINFECTION ON THE FORMATION OF DISINFECTION BY-PRODUCTS

N. Mole, M. Fielding and D. Lunt

WRc
Henley Road
Medmenham
SL7 2HD

1 INTRODUCTION

It has been suggested[1] that by-product formation is probably of minor importance in ultra-violet disinfection of water. However, in contrast, there have been a number of studies which indicated that under some conditions the formation of by-products could be of significance. For example, it has been reported[2] that, in the presence of nitrates or nitrites, elevated levels of mutagenic substances are formed from various amino acids on irradiation of water under neutral conditions. Further, it has been reported[3] that secondary amines in solutions containing nitrite can be photochemically converted to N-nitrosamines, that exhibit mutagenic and carcinogenic properties.

Experiments were carried out to investigate the influence of UV disinfection on the formation of disinfection by-products. These covered the formation of nitrite from nitrate, the formation of aldehydes, changes to the chlorination by-product potentials of humic acid solutions and river water, the formation of haloacetic acids from chlorinated solvents and Gas Chromatography-Mass Spectrometry (GC-MS) investigations of other possible by-products.

1.1 Formation of Aldehydes

Aldehydes can be formed from disinfection of potable water, particularly as a result of the use of ozonation. The major aldehydes produced from the ozonation of humic acid solutions are formaldehyde, acetaldehyde, glyoxal and methyl glyoxal, with the ozonation of river water giving rise to a similar range of compounds. Aldehyde production has been associated with taste and odour complaints and problems with bacterial regrowth in the distribution system. Therefore, the possibility of aldehyde production as a result of UV irradiation of humic acid and river water was investigated.

A published method[4] was used for the analysis of a range of aldehydes. Quantification was carried out of the aliphatic aldehydes C1-C10, glyoxal and methylglyoxal. Calibration standards were prepared by spiking deionised, activated-carbon-filtered water with a mixture of the aldehydes at concentrations covering the range 5-50 µg l^{-1}. These were then extracted and analysed in the same way as the samples.

A JaBay Model 2BM Ultraviolet flow-through reactor (Aztec Environmental Control, Didcot) with reactor volume 1 litre, and lamp power 20 W was used. Water was pumped through with a peristaltic pump. The lamp was allowed to warm up (with the reactor full of liquid) for 3 minutes before use. The UV dose applied, estimated by ferrioxalate actinometry, was 2.8 W l^{-1}. In order to evaluate the effect of higher UV doses in the flow-through system Thames river water was recirculated through the reactor; samples being taken at various times (see Table 1). These results demonstrate an increase in aldehyde formation with irradiation time, i.e. with increasing recirculation time. However, the levels formed are unlikely to be of any concern.

1.2 Effect of UV irradiation on trihalomethane formation potential and the formation of haloacetic acids

It has been found[5] that UV irradiation of an aqueous solution of humic acid (4.3 mg l^{-1}) for 15 minutes had only a minor effect on THM formation potential. To investigate the effect of UV irradiation on DBP precursors, a sample of Thames river water was chlorinated, after UV irradiation, under conditions designed to achieve DBP formation potentials. THM, trichloroacetic acid (TCA) and dichloroacetic acid (DCA) concentrations determined. THMs were determined by pentane extraction followed by quantification with Gas Chromatography-Electron Capture Detection (GC-ECD). DCA and TCA were determined by diethyl ether extraction at pH 0.5 followed by methylation and quantification with GC-ECD.

A 420 ml sample was irradiated in a batch photoreactor. The reactor was constructed from borosilicate glass with quartz-glass immersion and cooling tubes. UV irradiation was provided by a 15 W low pressure mercury lamp emitting at a wavelength of 254 nm. The applied UV dose was 2.6 W l^{-1}, as determined by ferrioxalate actinometry. Irradiation times of 10, 30, 60, 90 and 120 minutes were used, corresponding to UV doses of 1100, 3300, 6600, 9900, and 13000 mJ cm^{-2}. Aliquots of 400 ml were then chlorinated at 20°C for 17 hours (3.7 mg l^{-1} chlorine).

Table 1   *Aldehyde levels from Thames river water recirculated (4 l min^{-1}) through a continuous flow UV irradiation system*

Recirculation time min.	C1 µg l^{-1}	C2 µg l^{-1}	C3 µg l^{-1}	C4 µg l^{-1}	C9 µg l^{-1}	C10 µg l^{-1}	Gly µg l^{-1}	Me Gly µg l^{-1}
0	8.0	6.9	nd	nd	nd	0.7	nd	nd
5	15	13	0.6	0.5	nd	0.5	0.5	0.3
10	16	15	0.8	nd	nd	0.5	0.9	0.6
20	20	19	nd	nd	0.8	0.7	1.4	1.0
40	24	22	1.4	0.5	0.9	0.8	2.3	1.8
80	30	44	2.6	0.8	0.9	0.8	3.8	3.5
120	33	36	3.1	1.0	0.7	0.4	4.9	3.5
180	36	46	3.9	2.9	0.8	0.5	7.8	4.7

Cn	Aliphatic aldehyde with n carbons
Gly	Glyoxal
Me Gly	Methylglyoxal
nd	not detected

The concentrations of THMs, TCA and DCA were determined (i.e. the respective formation potentials); the results being shown in Table 2. Total THM formation potentials (THMFPs) decreased by 9% after UV irradiation at the highest dose used (13000 mJ cm^{-2}). There appeared to be an initial increase (after UV dose of 1100 mJ cm^{-2}) followed by a decrease (with increasing UV dose) in THMFP. The TCA formation potential does appear to decrease at the highest UV doses but does not show the initial increase observed for THMs. There is no evidence of significant variation in the DCA formation potential. These results do not correlate with the decrease in UV absorbance with increasing UV dose. Typically, UV absorbance is regarded as a measure of the reactivity of organic material, but this does not appear to be the case with respect to the formation of THMs, TCA and DCA.

In the JaBay flow through system decreases in levels of all the THMs were observed after UV irradiation with doses above 43 mJ cm^{-2}, with total THMs showing an 12% decrease after irradiation at the highest dose (105 mJ cm^{-2}). The results obtained are shown in Table 3. The increase in THM formation potentials observed for the lower doses in the batch reactor do not seem to have occurred in the flow through reactor. Since the doses applied using the flow through reactor are generally lower than those in the batch reactor it might have been expected that some increase in THMFP would have been observed.

Table 2   *Total THM and haloacetic acid formation potential in raw water chlorinated after UV irradiation in a batch UV photoreactor*

UV dose mJ cm^{-2}	UV absorbance at 254nm	Total THMFP µg l^{-1}	DCA µg l^{-1}	TCA µg l^{-1}
0	0.110	76	15	19
1100	0.092	87	17	21
3300	0.087	85	18	20
6600	0.086	70	16	16
9900	0.069	72	15	15
13000	0.062	69	14	14

THMFP = Trihalomethane formation potential

Table 3   *Total THMFP in raw water chlorinated after UV irradiation in a continuous flow UV irradiation unit*

UV dose mJ cm^{-2}	Total THMFP µg l^{-1}
0	83
40	84
43	82
59	79
80	76
105	73

THMFP = Trihalomethane formation potential

1.3 Irradiation of unsaturated chlorinated solvents

Chlorinated organic solvents such as trichloroethene (TCE) and tetrachloroethene (PCE) are common contaminants of groundwaters, especially in industrialised areas. The likely production of chlorinated by-products as a result of the irradiation of TCE and PCE with low (JaBay flow-through system) and high (batch UV photoreactor) UV doses was investigated.

Spiked solutions of TCE and PCE in borehole water were prepared at 100 and 1 000 µg l^{-1} by addition of methanolic solutions of the chlorinated solvents. The spiking volume used was 50 µl/l of water.

1.3.1 Batch photoreactor experiments. The UV lamp and immersion tube from the photoreactor described above were supported in a 2.5 litre capacity glass bottle with water circulating around the immersion tube to maintain constant temperature of 20°C. A two litre sample was placed in the bottle allowing enough headspace to ensure thorough mixing during irradiation periods of 2, 10, 30 and 60 minutes (corresponding to UV doses of 220, 1100, 3300 and 6600 mJ cm^{-2}). At the end of each irradiation period aliquots were taken for TCE, PCE and chlorinated acetic acids analysis. Results obtained are shown in Tables 4 and 5.

TCE removal was 42% and 30% for initial TCE concentrations of 100 and 1 000 µg l^{-1} respectively after 60 minutes irradiation (6600 mJ cm^{-2}). PCE removal was 93% for initial PCE concentrations of 100 and 1 000 µg l^{-1} after a 60 minute irradiation period. No production of either dichloroacetic acid (DCA) or trichloroacetic acid (TCA) was observed as a result of TCE degradation except at the highest UV doses and TCE concentration. Even under these conditions only small amounts of DCA and no TCA were detected.

The irradiation of PCE produced a UV dose dependent increase in DCA concentration. TCA was only formed in appreciable concentrations at the higher PCE concentration.

1.3.2 Continuous flow UV irradiation system. Five litre volumes of TCE and PCE at 100 and 1 000 µg l^{-1} in borehole water were prepared. These were passed through the reactor at flow rates of 4 l min^{-1} and 8 l min^{-1} (corresponding to UV doses of 120 and 60 mJ cm^{-2}). Aliquots were taken for analysis after approximately 4.5 litres had flowed through the reactor and been discarded. In order to estimate the effect of a higher UV dose in the flow-through system the TCE and PCE spiked solutions were recirculated through the reactor for a 30 minute interval at a flow rate of 4 l min^{-1} (UV dose 5700 mJ cm^{-2}). The results obtained are shown in Tables 6 and 7.

The removal of TCE and PCE at both spiking levels used was 20% or less with UV doses of 60 and 120 mJ cm^{-2}; an increase in irradiation time achieved by recirculation caused a 95% degradation of TCE and at least 99% for PCE (see Table 5). No DCA production was observed from TCE with any level of irradiation used in the flow-through system, while DCA levels with PCE irradiation increased with applied UV dose. TCA was generated only from irradiation of 1 000 µg l^{-1} PCE, but at significantly lower concentrations than DCA.

The results from these two sets of experiments indicate that the use of UV irradiation in the treatment of PCE contaminated waters has the potential to result in the formation of appreciable concentrations of DCA and, to a much lesser extent, TCA. The formation of TCA requires a molecular rearrangement and is thus less kinetically-favoured than is the formation of DCA.

Table 4   *Effect of UV irradiation on trichloroethene (TCE) spiked ground water (100 and 1 000 µg l^{-1})*

UV dose mJ cm^{-2}	TCE µg l^{-1}	DCA µg l^{-1}	TCA µg l^{-1}
0	105	<0.5	<0.5
220	93.1	<0.5	<0.5
1100	87.7	<0.5	<0.5
3300	84.8	<0.5	<0.5
6600	63.1	<0.5	<0.5
0	994	<0.5	<0.5
220	829	<0.5	<0.5
1100	796	0.7	<0.5
3300	822	0.6	<0.5
6600	707	1.2	<0.5

Table 5   *Effect of UV irradiation on tetrachloroethene (PCE) in spiked ground water (100 and 1 000 µg l^{-1})*

UV dose mJ cm^{-2}	PCE µg l^{-1}	DCA µg l^{-1}	TCA µg l^{-1}
0	108	<0.5	<0.5
220	89.8	2.4	<0.5
1100	61.0	9.8	0.9
3300	20.4	13.0	0.9
6600	7.3	6.0	-
0	1110	<0.5	<0.5
220	413	24.3	4.6
1100	658	79.0	7.6
3300	237	153	15.7
6600	68.6	180	53.0

Table 6   *Effect of UV irradiation in the JaBay flow-through system on trichloroethene spiked ground water (100 and 1 000 µg l^{-1})*

UV dose mJ cm^{-2}	TCE µg l^{-1}	DCA µg l^{-1}	TCA µg l^{-1}
0	106	<1.0	<1.0
60	90.0	<1.0	<1.0
120	93.6	<1.0	<1.0
5700	5.3	<1.0	<1.0
0	990	<1.0	<1.0
60	930	<1.0	<1.0
120	948	<1.0	<1.0
5700	46.0	<1.0	<1.0

Table 7 *Effect of UV irradiation in the JaBay flow-through system on tetrachloroethene spiked ground water (100 and 1 000 µg l^{-1})*

UV dose mJ cm^{-2}	PCE µg l^{-1}	DCA µg l^{-1}	TCA µg l^{-1}
0	101	<1.0	<1.0
60	89.6	4.4	<1.0
120	80.7	7.4	<1.0
5700	<1.0	9.0	<1.0
0	981	<1.0	<1.0
60	810	27.8	4.1
120	837	40.2	2.0
5700	<10.0	194	11.2

1.4 Formation of nitrite by UV irradiation of nitrate-containing water

It has been reported[6] that the photolysis of 50 mg l^{-1} nitrate yielded only 1.8 µg l^{-1} nitrite with irradiation close to what the authors considered the optimal for bactericidal efficiency, i.e. the nitrite concentration was well below the UK drinking water limit of 100 µg l^{-1}.

A set of experiments was carried out to investigate the formation of nitrite from the irradiation of nitrate-containing water. This study used the flow-through reactor in which the pH of de-ionised water was adjusted using sodium hydroxide and hydrochloric acid, as appropriate. The pH adjusted waters were spiked with 50 mg l^{-1} nitrate and subjected to UV doses of 60 and 120 mJ cm^{-2}. The samples were also recirculated through the reactor for 30 minutes, giving a UV dose of 5700 mJ cm^{-2}. The results obtained are given in Table 8. Nitrite was determined spectrophotometrically by the official Standing Committee of Analysts method (1981).

With one exception, nitrite was only detected in the recirculated samples. This suggests that, if the UV dose applied is close to the minimum recommended by the World Health Organisation for adequate disinfection (16 mJ cm^{-2}), then the UK drinking water limit should not be exceeded.

1.5 Investigation into the possible formation of nitrogen-containing organics on irradiation of water containing nitrate

To investigate whether the irradiation of waters containing nitrate can give rise to the formation of nitrogen-containing organics, a screening study was carried out. Two litre volumes of Thames river water were spiked with 50 mg l^{-1} nitrate and irradiated for 1 hour in the batch reactor (UV dose 6600 mJ cm^{-2}). Borehole water was also spiked and similarly irradiated. Control samples of water without nitrate added, and without irradiation were also generated. One litre aliquots were saturated with 25 g sodium chloride and extracted with diethyl ether (350 cm^3) after the pH was adjusted to 2.0. The pH of the residual aqueous phase was then adjusted to 12.5 with 10M sodium hydroxide, followed by extraction with 100 cm^3 of diethyl ether. Residual water was frozen out of the ether extracts, and the ether decanted off and concentrated to approximately 3 cm^3 by

Table 8   *Effect of UV irradiation in the JaBay flow-through system on nitrite levels in nitrate (50 mg l^{-1}) spiked deionised water*

pH	UV dose mJ cm^{-2}	Nitrite µg l^{-1}
6	0	<33
	60	<33
	120	<33
recirculated 30 minutes	5700	790
7	0	<33
	60	<33
	120	<33
recirculated 30 minutes	5700	520
8	0	<33
	60	<33
	120	120
recirculated 30 minutes	5700	550

Kuderna-Danish evaporation. The final volume was adjusted to 2 cm^3 using nitrogen blow down.

The concentrated samples were analysed by GC-NPD (nitrogen/phosphorus detector) using a DB5.625 capillary column; the initial column temperature was 30°C (held for 2 minutes) then increased by 7°C min^{-1} until it reached 200°C. Dichloroacetonitrile (DCAN) and nitrobenzene (both at 10 µg cm^{-3}) were used to check the detector response.

Chromatographic differences were seen between the irradiated and non-irradiated samples at both extraction pH values, particularly for the river water samples. The irradiation of river water produced larger peaks at retention times 15.25 and 19.64 minutes than were present in the unirradiated river water. The addition of nitrate, followed by irradiation produced further enhancement of these peaks, and generated a set of peaks, retention time 24.7-25.6 minutes. Whilst the use of NPD meant that no distinction could be drawn between nitrogen and phosphorus containing compounds, the change on the addition of nitrate indicated that nitrogen-containing compounds may have been formed.

These peaks were further investigated by GC-MS. The analyses of the nitrate-spiked river water samples did not correlate with the results of the GC-NPD screening experiment. The UV irradiation of the river water with no added nitrate (although containing about 40 mg l^{-1} of naturally-occurring nitrate) produced a number of aldehydes, including hexanal and benzaldehyde, and some unknown compounds tentatively identified as unsaturated amides (C_{16} and C_{18}). However, the unknown amides did not seem to be present in the UV irradiated samples with added nitrate. The non-irradiated samples with added nitrate contained a number of possible nitrogen-containing compounds, none of which were found in the irradiated sample. There is a possibility that contamination was introduced when the nitrate was spiked. There was no evidence of phosphorus-containing by-products.

A more detailed study was carried out, in which levels of added nitrate and nitrite were varied and samples extracted at acid and alkaline pH. Where UV irradiation was used the dose applied was 6600 mJ cm^{-2}. This more detailed study showed that the solution of nitrate used did not contain any nitrogen-containing organic compounds. However, the other results obtained in the initial study were not confirmed. Two samples were identified as containing possible nitrogen-containing organic compounds. These were the river water samples with added nitrite which received no UV irradiation and were extracted at pH 2. Of the conditions used these were those most likely to give rise to nitrosamines. However, the peaks observed could not be specifically identified, and when UV irradiation was applied the peaks were not observed. There was no evidence of phosphorus-containing by-products.

1.6 Summary

The results obtained can be summarised as follows:
- from the results obtained in this study, if UV units are operated at a dose of around 100 mJ cm^{-2}, then at high nitrate levels the concentration of nitrite may approach or exceed the regulatory limit of 100 µg l^{-1};
- at low UV doses (i.e. 25-30 mJ cm^{-2}), pH values typical of UK water sources, and nitrate concentrations within the regulatory limit, the formation of nitrite is rarely, if ever, likely to exceed 100 µg l^{-1};
- aldehydes, in particular formaldehyde and acetaldehyde, are formed as a result of the UV irradiation of Thames river water. However, they do not appear to be formed at levels that would be of concern;
- UV irradiation may reduce the chlorination by-product formation potential (THMs and TCA) of humic acid solutions and Thames river water, but the changes observed are too small to be categorical. Under some conditions there may be a slight increase in formation potential;
- the use of UV irradiation to treat water contaminated with PCE can give rise to significant concentrations of DCA. TCA can also be formed but generally at lower concentrations. This may also be a consideration where UV-based advanced oxidation processes are used to remove PCE;
- from the results obtained in this study, if UV units are operated at a dose of around 100 mJ cm^{-2}, then if PCE is present in high concentrations, i.e. at, or above, 1 mg l^{-1}, then the concentrations of DCA produced may approach the WHO Drinking Water Guideline value of 50 µg l^{-1}.
- UV irradiation of nitrate-containing river water results in the apparent formation of nitrogen- or phosphorus-containing compounds. However, GC-MS experiments failed to confirm the formation of any compounds of interest.

References

1. R.L. Wolfe, *ES and T*, 1990, **24**, 768.
2. J. Suzuki, T. Ueki, S. Shimizu, K. Uesugi, and S. Suzuki, *Chemosphere*, 1985, **14**, 493.
3. T. Ohta, J. Suzuki, Y. Iwano and S. Suzuki, *Chemosphere*, 1982, **11**, 797.
4. M.J. Sclimenti, S.W. Krasner, W.H. Glaze and H.S. Weinberg, *Proceedings of AWWA WQTC*, San Diego, Ca., 1991, 471.
5. R.A. Sierka and G.L. Amy, *OS and E*, 1985, **7**, 47.
6. C. von Sonntag and H-P. Schuchmann, *J. of WSRT*, 1992, **41**, 67.
7. WHO, 'Guidelines for Drinking Water Quality. Second Edition. Volume 2 Health Criteria and Supporting Documentation', World Health Organisation, Geneva, 1996.

IMPACT OF PRE-OZONATION AND SLOW SAND FILTRATION ON DISINFECTION BY-PRODUCTS

C. J. Cable

North West Water
Technology Development Team
Huntington Water Treatment Works
Chester, CH3 6EA

1 INTRODUCTION

1.1 Oswestry Water Treatment Works

North West Water's Oswestry Water Treatment Works was constructed over one hundred years ago to treat water from the newly constructed Lake Vyrnwy Reservoir in North Wales in order to augment the existing water supply to the Merseyside area. Treatment was, until recently, by slow sand filtration, chlorination and pH adjustment. After additional phases of construction in this century, the maximum treatment capacity is now 205 Mld^{-1}.

As is typical with water impounded in upland reservoirs, the water quality suffers from seasonal discoloration owing to material of humic origin, and the single stage of slow sand filtration has not always been sufficient to reduce this colour to levels acceptable to NWW's customers. A previous paper outlines pilot plant studies of a pre-ozonation stage retrofit at Oswestry.[1] The trials demonstrated that pre-ozonation was a viable option for the control of colour and, accordingly, a full-scale Trailligaz ozone plant was built; this first went into service in May 1994. Since then, experience with the full-scale plant has fully vindicated the decision to install ozonation. In order also to address THMs, the ozone plant is currently operated more frequently than strictly necessary to control colour.

1.2 Ozone for Removal of THMs

1.2.1 The THM issue. Also associated with the coloured organic material in the raw water, there is an issue with trihalomethanes (THMs) in the distribution zones served by Oswestry — concentrations approaching the 100 μg l^{-1} Prescribed Concentration or Value (PVC) have been detected at the customer's tap on occasions. The limitation of slow sand filtration in removing organic THM precursor material is evident, and this is compounded by the lengthy residence times in the supply aqueduct, service reservoirs and distribution infrastructure, and in the application of rechlorination at various points along the way.

1.2.2 Mechanism for the discharge of colour. The mechanism for removal of extra humic colour by ozone is thought to be largely a one-stage process, involving alteration of the nature of chemical bonding in the humic macromolecules. The result is less extensive conjugation of unsaturated groups — the property that causes the organic matter to be coloured. Probably, very little destruction by mineralisation of the organic matter takes place during the ozonation stage. Comparatively little further reduction in colour is then anticipated over the biological filtration stage (Figure 1A).

1.2.3 Mechanism for removal of THM Precursor Material. An enhancement in the removal of the organic THM precursor material, on the other hand, is anticipated because of a quite different mechanism involving a synergistic two-stage process of ozonation followed by slow sand filtration (Figure 1B).[2,3] The ozone-transformed organic matter arriving on the slow sand filter is theoretically more susceptible to biological degradation on the filter's biological layer or *schmutzedecke*. The increased loading of biodegradable organic matter should, in turn, activate and increase the biological community in the *schmutzedecke*, further increasing the removal rate.

1.2.4 Anticipated Benefits. This can be expected to result in a lowering of the tendency to form THMs during primary disinfection and in distribution in two ways:
- directly, because of the lower concentrations of THM precursor material available to interact with chlorine.
- indirectly, because the lower concentrations of total organic matter should reduce the chlorine demand of the filtrate, enabling the chlorine dosing rate to be lowered.

Figure 1. *The likely fate of organic matter when subject to: A) aeration followed by biological filtration and B) ozonation followed by biological filtration.*

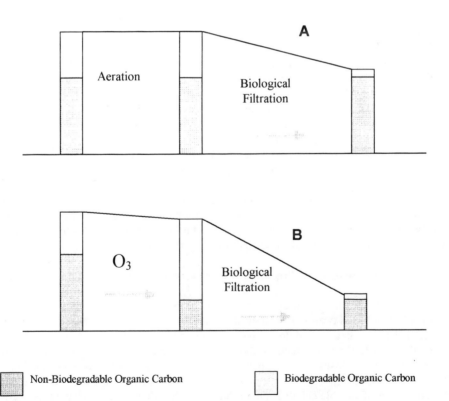

This rationale depends on the formation of THMs being directly proportional to the level of organic matter in the treated water, regardless of the nature of this organic matter, e.g. whether on not converted by ozonation. The physical situation is likely to be more complicated than this.

This potential additional benefit of installing ozonation at Oswestry was not fully explored during the original pilot trials; and, indeed, the full-scale plant was justified solely on the grounds of the demonstrated colour reduction.

1.3 Objectives

It was the intention of the work described in this paper, firstly to demonstrate whether pre-ozonation at Oswestry does lead to a significant reduction in THM formation potential. This was assessed by analysis of present and past data over the years preceding and following commissioning of the ozone plant. A short three-week trial has also been carried out. The aim of this was to find out whether an increase in ozone dose could provide a partial or complete solution to the THM concerns in the Oswestry distribution zones. A suitable operating strategy to maximise the reduction of THMs would almost certainly involve operating the ozone plant at higher power levels than is the case in the present strategy to control colour.

2 RESULTS

2.1 Analytical Methods

2.1.1 Trihalomethane Formation potential. THM Formation potentials (THM FPs) were carried out according to North West Water's Quality Instruction Document, which was in, turn based on the AWWA Standard Method.[4] The results are quoted in the text in terms of $\mu g l^{-1}$ of total THMs. The method involves chlorinating one litre samples with a standard excess of chlorine in the dark at standard temperature for seven days. The THMs in the resulting solution are then determined.

2.1.2 Colour Determinations. True colours quoted in °H were measured using a UV absorption method at 400 nm after filtration of the samples through a 4.5 μm porosity membrane.

2.1.3 Other Determinations. These and all other determinations — Total Organic Carbon (TOC), Turbidity, Trihalomethane Concentrations — were all carried out using the NWW Quality Instruction Documents, based in turn on Methods for the Examination of Water and Associated Materials series published by HMSO.

2.2 Demonstration of THM Precursor Material Removal Over Three Year Period

2.2.1 Details of the works. Oswestry Water Treatment Works (shown in diagrammatic form in Figure 2) has twenty-three slow sand filters of 55 m x 62 m dimensions all open to the air and each with a normal flow capacity of around 9 Mld^{-1}. Owing to the desirability of using the head at Vyrnwy Aqueduct above the works to power a generator, one stream of water normally by-passes the ozone plant. Hence a lane of six filters (the second column of filters from the left in Figure 2) receive water which has not normally been ozonated. Filtered water of suitable colour is then achieved by blending. The ozone plant is operated at levels varying between 1.5 and 3.2 mg l^{-1} O$_3$, depending on the quality of raw water.

2.2.2 The Sampling Programme. Samples of raw water for the ozonated stream were taken from the intake at Llanforda Reservoir. Samples of raw water for the non-ozonated stream were taken from the generator house. The clearwell outlets of Filters 9 and 11 (Figure 2) were sample points for the filtered water from the ozonated and non-ozonated streams respectively. Sample batches were collected at various times of year representing a range of raw water qualities and ozone plant operating levels. Parameters measured included trihalomethane formation potential (THM FP) true colour, total organic carbon (TOC) and turbidity, all of which are considered to be good indicators of THM precursor material.

2.2.3 Findings. The trial covered a period of two years, from September 1994 to August 1996. The average results for colour, TOC and THM Formation Potential at the four sample points are listed in Table 1. In addition, the THM FPs for each sample batch are summarised in Figure 3.

In general, the values for colour, TOC and THM Formation Potential were marginally higher at the ozonated stream raw water sample point than at the non-ozonated stream raw water sample point. This was not unexpected, as Llanforda Reservoir has its own catchment, which apparently contributes some additional organic colour to the water

Figure 2. *Diagrammatic representation on Oswestry WTW filters, showing Filter 9 (ozonated stream) and Filter 11 (non-ozonated stream).*

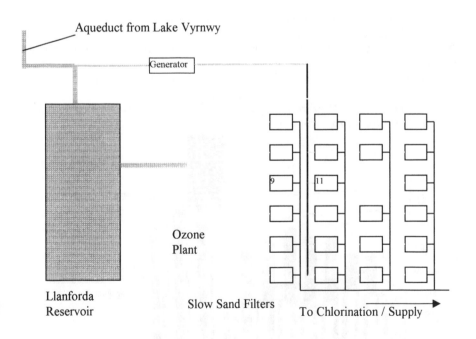

arriving from Vyrnwy Aqueduct. Fortunately, the difference in raw water quality was not significant enough to invalidate the results.

Table 1 *Summary of Results for the Four Sample Points.*

Determin-and	Unit	O_3 Raw	Non-O_3 Raw	Filter 9	Filter 11	% Removal O_3	% Removal Non-O_3
Colour	°H	21.9	21.6	11.7	17.6	46.4	18.5
TOC	mgCl⁻¹	3.82	3.63	2.71	3.08	29.1	15.1
THM FP	µgl⁻¹	439	428	220	320	49.8	25.3

The enhancement in the removal of colour by ozone was the most pronounced. This is in accord with the one stage denaturing scenario for organic molecules by ozone, outlined above. However, removal of THM Formation Potential was also significantly enhanced in the ozonated stream. Levels of all three determinands were consistently lower in the ozonated water in all the sample batches. This indicates that ozone does have a role in reducing THMs in distribution.

2.3 THM Concentrations in Distribution

2.3.1 Tracing the Origin of Water in Distribution. Treated water from Oswestry Water Treatment Works serves Merseyside plus additional regions in the south of NWW's distribution area. The distribution zones are very diverse in terms of transport time down the aqueduct, local residence times in the distribution network, and number of rechlorinations that the water is subject to. Additionally, there are some mixed zones,

Figure 3 *Summary of THM FP removal rates at Oswestry WTW: batches 1 to 8 taken Mar to Oct; batches 9,10,11 taken Nov to Feb all during 1994, 1995 and 1996*

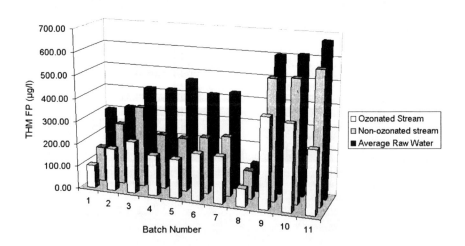

served also by water from lowland river sources (which has been treated by coagulation-sedimentation and rapid gravity filtration) and from boreholes. The composition of the water in these mixed zones will often vary on a temporal basis. Fortunately, Oswestry final water is of relatively low electrical conductivity (around 65mScm^{-1}) while that from lowland river sources has a much higher conductivity. Hence, it is possible to trace the water's origin.

2.3.2 THMs in Distribution. There are a small number of these zones giving rise to concerns regarding THMs. However, they are zones in which Oswestry water predominates and the other sources do not appear to be implicated. As expected, the troublesome zones have comparatively long water residence times.

By looking at the period before and after commissioning of the Oswestry ozone plant in 1994, it was evident that the ozone had had some mitigating effect on the THM concentrations in the key zones. However, the raw water was of unusually low organic content during the eighteen months following the drought of 1995, and this may have distorted the results to some extent. It was indicated that the current operation of the ozone plant may not be appropriate to ensure THM compliance in all cases.

2.4 Elevated Ozone Dose Short Duration Trial

During summer 1987, starting on September 1st, all the raw water — rather than the usual three-quarters — was routed through the ozone plant for a period of three weeks. The ongoing set point of ozonation rate per unit volume (1.7 mgl^{-1}) was maintained during the first two weeks of this period of increased flow. During the third week (15th Sept – 22nd Sept) the ozonation rate per unit volume was raised to 2.7 mgl^{-1}, still treating the whole flow. THM FP, TOC and colour were measured at the raw water sample point and at the clearwells of Filters 9 and 11 twice a week during this period and during an initial baseline period. THM concentrations at a set point along the Vyrnwy Aqueduct were measured daily. The relationship observed between ozonation rate per unit volume and total THM concentration at the set point down Vynwy Aqueduct is shown in Figure 4.

During the two weeks following the 1st Sept. (with all the water routed through the ozone plant), the THM concentrations at the fixed sample point down the supply aqueduct did not significantly improve. However, when the ozonation rate was raised from 1.7 mgl^{-1} to 2.7 mgl^{-1} on the 15th Sept, there was a significant reduction in THM concentrations. These findings were supported by the other water quality data.

Almost immediately after the ozone dosing was applied to Filter 11, this filter showed an increased performance in terms of removal of colour, TOC and THM FP, nearly matching the performance of Filter 9. Hence, a lengthy period of filter acclimation did not appear to be necessary to achieve enhanced filter performance when introducing pre-ozonation. The lower than expected reduction of THMs during the first two weeks of the trial was probably owing to the reduced effectiveness of the ozone plant at transforming organic material at lower contact times. However, it is not clear whether the raw water colour remained constant over the period of the trial (Figure 4).

3 CONCLUSIONS

3.1 Summary of Findings

The first part of the study clearly proved that ozonation combined with slow sand filtration

Figure 4. *Ozone Plant Trials*

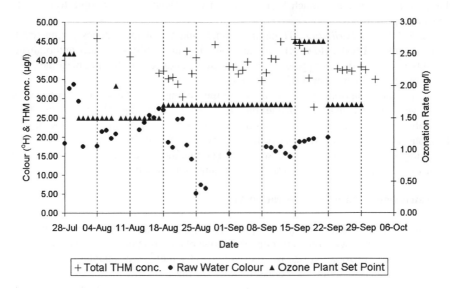

does have a significant effect on reducing THMs in distribution. This conclusion is supported by the archive data in the distribution zones before and after the ozone plant was commissioned.

There appeared to be a reduction in THM with increased ozone dose during the three-week plant trial. The evidence for this is far from conclusive, however, and a longer trial is needed.

3.2 Proposed Future Work

It is proposed to run the ozone plant at a high rate through a prolonged period of highly coloured raw water input. Monitoring of the chlorine demand might enable the chlorine dosing to be reduced in line with the enhanced removal of organic matter. Hence two of the factors leading to reduced production of THMs will be acting concertedly. A reliable raw water quality trigger is clearly needed to determine the operating ozonation rate — currently, the ozonation rate for colour removal is determined by final (blended) water quality, especially colour. Raw water colour is perhaps the most promising determinand. The results of the three-week trial suggest that the best approach to improve THM FP reduction is to run three-quarters of the water through the ozone plant and continue to achieve the final water quality by blending ozonated and non-ozonated streams.

Much work has been carried out on the mechanisms by which ozone transforms organic molecules.[5] Molecular addition and free radical attack are both known to be important. And the way that ozone degrades organic matter is likely to be highly dependant on such factors as the alkalinity, carbonate chemistry and pH. Some work along these lines, to explore the interaction of lime, CO_2 and O_3, has been carried out at full-scale at a plant in France.[6] It is, therefore, proposed to initiate work at pilot-plant scale, in order to maximise reduction of THM Formation Potential by chemical dosing, ozonation and biological filtration.

References

1. Greaves, P. G. Grundy and G. S. Taylor, 'Slow Sand Filtration: Recent Developments in Water Treatment Technology', N.J.D. Graham (Ed), Ellis Horwood Ltd, Chichester, 1988, Chapter 2, Section 2.4, p. 153.
2. Ferguson, J. T. Gramith and M. J. McGuire, *Jour. AWWA*, 1991, 85, N° 5, 32.
3. Singer and S. Chang, 'Impact of Ozone on the Removal of Particles, TOC, and THM Precursors', American Water Works Association Research Foundation and American Water Works Association, Denver, Colorado, USA, 1989.
4. AWWA, Trihalomethane Formation Potential; AWWA. 'Standard Method for Examination of Water and Wastewater', 17th Edition, 1989.
5. Langlais, D. A. Reckhow, and D. R. Brink (Ed) 'Ozone in Water Treatment: Application and Engineering.' Cooperative Research Report, American Water Works Association Research Foundation and Compagnie Général des Eaux', Lewis Publishes, Michigan, USA, 1991.
6. Pailhard, B. Legube, M. M. Boubigot and E. Leefebvre, *Ozone Sci. Engrg.,* 1989, 11, 93.

THE EFFECT OF NANOFILTRATION ON THE TRIHALOMETHANE FORMATION POTENTIAL OF COLOURED UPLAND WATERS

John W Littlejohn, Water Quality and Regulation Manager

North of Scotland Water Authority
Turriff Treatment Works, Shandscross
Turriff, AB53 7PL

1 INTRODUCTION

The North of Scotland Water Authority supplies 424 mega-litres of water to 1.2 million customers living in an area that extends over some 60% of the area of mainland Scotland, and also includes the islands of Orkney, Shetland and the Western Isles.

The Authority operates 305 water treatment works, of which 284 have individual capacities of less than 3 mega-litres per day. The treated water is distributed by way of a pipeline network comprising around 20,000km of water mains and over 700 service reservoirs, 91% of which have capacities of less than 2 mega-litres.

In 1997 there were 299 water supply zones within the Authority's area, but 250 of these zones served populations of less than 5,000, with a significant number serving populations of less than 100, some even as small as 10.

Although the larger regional networks are served with relatively modern two or three stage treatment works the treatment at the small supplies consists generally of simple up-flow filtration and chlorination. Many of these small supplies draw their water from peaty coloured upland lochs, streams and reservoirs where water quality is often very variable and weather dependant. Significant exceedances of Water Regulation Standards occur for colour, iron, oxidisability and total trihalomethanes.

Since taking over operational control of these supplies following local government re-organisation on 1 April 1996 the North of Scotland Water Authority has been developing a strategy for up-grading all poorly treated supplies. An important part of this strategy is the use of membrane technology to remove contaminants from the water, especially bacteria, iron, colour, and other organic molecules. Five such plants currently operate within the Authority's area (Table 1), with another currently undergoing commissioning.

With the exception of the Armadale works all the membranes are constructed from organic polymers. At Armadale the membrane is ceramic in nature.

This paper will deal with data relating to the Tomnavoulin and Bracadale water treatment works.

2 TOMNAVOULIN

Tomnavoulin is a village lying some 30 miles south of Elgin. A population of 120 people require a water supply of 60m^3 per day. This water is abstracted from a small stream which flows off an area of mainly heather covered hillside, although a little afforestation and agricultural activity also takes place in the catchment area.

The quality of the abstracted water is summarised in Table 2.

Prior to the introduction of the membrane plant, the water was treated by simple pressure filtration followed by chlorination to a free chlorine residual of 0.2 to 0.4mg/l after roughly 24 hours contact in the final water tank. Chlorine doses applied varied depending on the quality of the abstracted water, but generally lay in the range 1 to 4mg/l chlorine. Total trihalomethane (THM) levels also fluctuated with water quality, but usually fell into the range 50-100μg/l, with occasional samples exceeding 100μg/l (Figure 1).

With the introduction of the membrane plant during 1995 the colour of the water supplied to the customers reduced immediately to 5 Hazen or less and has remained at those levels to the present time (Figure 2). The target free chlorine residual was maintained at 0.2 to 0.4mg/l after storage, however the chlorine dose necessary to achieve such a residual fell to less than 1mg/l and currently averages about 0.8mg/l chlorine. Total Organic Carbon (TOC) concentrations did fall when the plant was introduced, but not by as much as was expected, staying around 1 to 2 mg/l Carbon (Figure 3). It would appear that it was only when the original membrane was replaced with a new membrane during 1996, following marked membrane surface deterioration, that TOC levels fell to below 1mg/l Carbon. Recently it has been noted that TOC concentrations have risen back into the 1 to 2 mg/l range.

Total trihalomethane results tend to mirror the TOC figures (Figure 1) with a fall from an average of about 65μg/l to around 35μg/l following the introduction of the membrane plant. When the membrane was replaced there was a further improvement in THM concentration, with the mean falling to about 18μg/l, but again recent figures have shown an upward turn similar to the TOC results.

Table 1 *Membrane Water Treatment Works*

Membrane Works	Volume Treated m³/day	Date into Service	Filtration Type
Tomnavoulin	60	October 1995	Nanofiltration
Kirkmichael	60	October 1995	Nanofiltration
Crathie	20	July 1996	Microfiltration
Bracadale	100	November 1996	Nanofiltration
Armadale	20	January 1998	Nanofiltration
Backies	1600	Being commissioned	Nanofiltration

Figure 1   *Tomnavoulin Water Supply - Total Trihalomethanes*

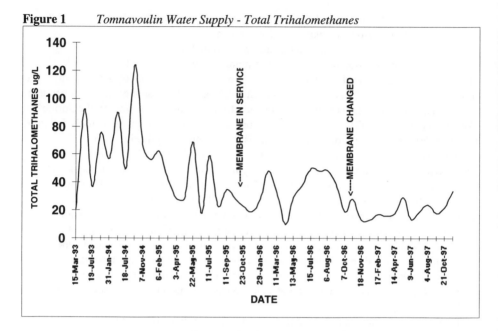

Figure 2   *Tomnavoulin Water Treatment Works - Colour Hazen*

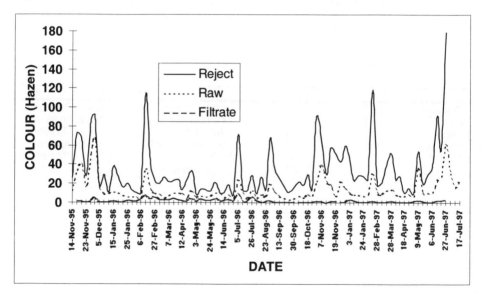

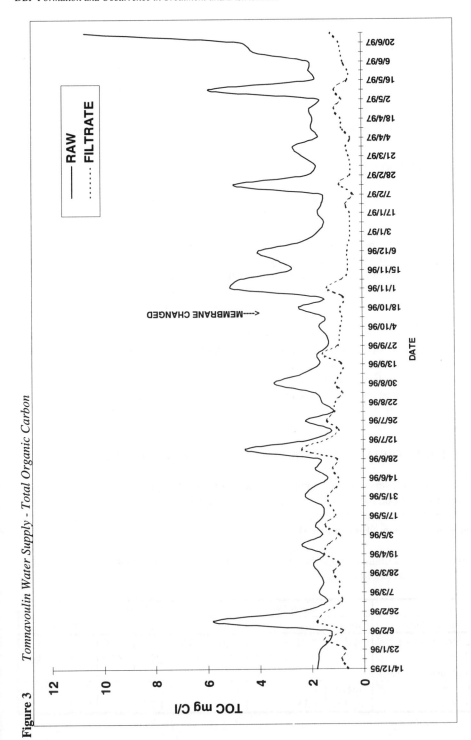

Figure 3 *Tomnavoulin Water Supply - Total Organic Carbon*

Table 2 *Tomnavoulin Water Treatment Works - Quality of Water Abstracted for Supply*

Parameter	Units	Typical Range
Total Coliforms	per 100ml	0 to over 500
Colour	Hazen	5 to 70
Iron	μg/l Fe	20 to 800
Total Organic Carbon (TOC)	mg/l C	1 to 10

3 BRACADALE

Bracadale is a community on the west coast of Skye. A population of 250 is supplied with $100m^3$ per day of water from the Voaker Burn. Prior to the introduction of the membrane plant during November 1996 the water was only treated by disinfection with chlorine before passing to supply. As the water colour varied from less than 20 Hazen to over 300 Hazen (Figure 4) the chlorine dose required to give adequate disinfection ranged from 2 to over 10mg/l chlorine, and the difficulties in controlling the dosing resulted in residual chlorine levels at the customers' tap, ranging from zero to over 1mg/l. The THM levels reflected these disinfection problems with concentrations varying from zero to nearly 500μg/l (Figure 5).

Since the introduction of the membrane plant the colour of the treated water has fallen to 10 Hazen or less (Figure 4), and THM levels to under 20μg/l. TOC levels in excess of 10mg/l carbon are being reduced to less than 3mg/l and often to between 0.5 and 1.0 mg/l carbon. Chlorine dosing has fallen to less than 1mg/l chlorine, although this requires to be adjusted upward as the residual is currently being lost too early in distribution. There was a brief period of higher THM levels during June 1997, with results up to 50μg/l. The reason for this is not known.

4 TOTAL TRIHALOMETHANE FORMATION POTENTIAL TESTS (TTFP)

Because THM levels at Tomnavoulin remained higher than originally expected TTFP tests were carried out on the raw and filtered water to compare the THM concentrations produced by both waters (Table 3). Similar tests were carried out on samples of water from the Bracadale plant (Table 4 and 5).

Table 3 *Total Trihalomethane Formation Potential Tests - Tomnavoulin Membrane Plant 24 hours Contact Time*

Date		30/09/96				08/11/96			
Sample Point	Units	Raw		Filtered		Raw		Filtered	
Chlorine Dose	mg/l	1.0	3.0	1.0	3.0	1.0	3.0	1.0	3.0
Total Trihalomethane	μg/l	55.4	77.4	23.5	52.1	29.5	87.4	18.3	24.5
Colour	Hazen	4		1		23		2	
TOC	mg/l C	1.35		-		4.75		0.97	

Figure 4   *Bracadale Water Supply - Colour in Distribution*

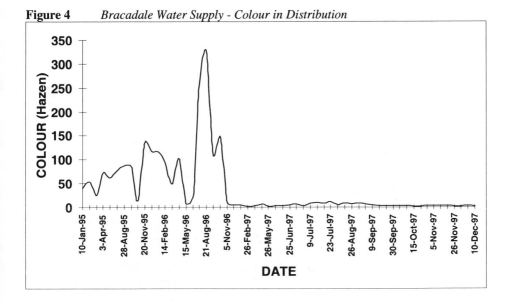

Figure 5   *Bracadale Water Supply - Total Trihalomethanes in Distribution*

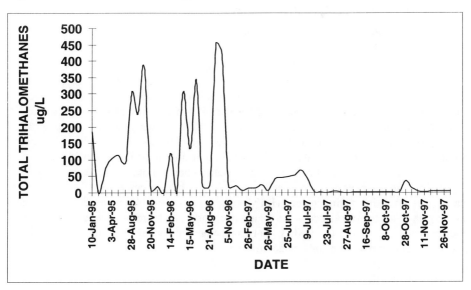

Table 4 *Total Trihalomethane Formation Potential - Bracadale Water Supply - 16/9/96*

Chlorine Dose (mg/l Cl)	Parameter	Raw Water Colour = 90 Hazen TOC = 11mg/l C	Filtered Water Colour = 8 Hazen TOC = 3mg/l
		24 hr Contact	24 hr Contact
0.5	Total Trihalomethane μg/l	5	14
1.0	Total Trihalomethane μg/l	21	45
3.0	Total Trihalomethane μg/l	88	98

The results from both locations indicate that it is possible, under certain high chlorine dose conditions, to produce quite high THM concentrations in water filtered through a membrane filter. Equally, it is clear that this does not occur every time the water is highly chlorinated. Fortunately chlorine doses in excess of those necessary to maintain an adequate chlorine residual in these small supplies are required before any truly significant THM concentrations are reached. Season of the year, and the effect that may have on the nature of organics present, condition of the membrane, the nature of the raw water catchment and the type of membrane all could potentially influence THM formation in the filtered water. Further investigation of this phenomenon is required, especially if more widespread use of membrane technology is envisaged. For example the use of lowland water sources, or the need for secondary disinfection in larger distribution systems, may result in significant levels of THMs being produced.

5 COMPARISON WITH CHEMICAL TREATMENT

Because these supplies are so small treatment by coagulation is not generally a viable option, therefore little data exists which one can use to compare THM levels produced by conventional coagulation and filtration with those produced following nanofiltration. However across the Authority's area there are a handful of water treatment works where one might use the data to make such a comparison of THM concentrations. These works, treating quite small volumes of highly coloured raw water by coagulation with aluminium sulphate, produce water where the THM concentrations fall in the range 4 to 98μg/l (Table 6). The works giving the higher concentrations being those where traditionally higher chlorine residuals are maintained, circa 0.5mg/l chlorine compared with 0.1mg/l chlorine at those sites giving lower THM values. These lower chlorine residuals are more akin to the residuals maintained at the membrane works and therefore it would appear that the THM concentrations produced by the two treatment methods are, in fact, very similar, given similar chlorine dose and contact time.

6 CONCLUSION

The introduction of nanofiltration to these highly coloured water has resulted in the supply of clear, colourless water where, in the past, this had not been possible, with an associated reduction in the TOC concentration of about 70 to 90+%. With this reduction in organic loading has come a significant fall in the chlorine demand of the water, with chlorine dose levels dropping at times by several milligrams per litre.

Table 5 *Total Trihalomethane Formation Potential - Bracadale Water Supply - 3/12/96*

Chlorine Dose (mg/l Cl)	Parameter	Raw Water (Colour = 30 Hazen TOC = 4.29mg/l C)			Filtered Water (Colour = 3 Hazen TOC = 0.99mg/l C)		
		1 hr Contact	24 hr Contact	72 hr Contact	1 hr Contact	24 hr Contact	72 hr Contact
0.5	FCR	-	-	0.00	-	-	0.02
	THM	-	-	12	-	-	23
1.0	FCR	-	0.00	0.00	-	0.88	0.70
	THM	-	24	37	-	21	26
3.8	FCR	1.36	0.20	Trace	3.44	3.20	2.90
	THM	83	131	182	14	27	32

FCR = Free Chlorine Residual (mg/l Cl)   **THM = Total Trihalomethane (µg/l)**

Table 6 *Total Trihalomethane Concentrations at Small Supplies treated by Chemical Coagulation*

Water Treatment Works	Total Trihalomethane Concentrations in Distribution 1996 - µg/l		
	Minimum	Mean	Maximum
Sandwick	31	61	91
Bressay	17	52	98
Cullivoe	36	46	52
Lairg	17	22	32
Clunas	23	45	80
Sandy Loch	33	56	72
Tolsta	4	9	12
Stornoway	5	10	17

The overall effect of this has been a reduction in THM levels which, when the water is highly coloured, has been in the order of 95+%, from near 500µg/l to less than 20µg/l. When compared to similar waters treated by chemical coagulation the THM concentrations produced by the membrane treated waters are generally lower than those in the chemically treated water. Chlorination practice at each individual site influences THM levels obtained and this may account for the differences in THM concentration between the sites rather than the method used for colour removal.

However there are occasions when the THM concentrations in membrane filtered water rise above background levels. The reasons for this are unclear but may be related to membrane condition and/or seasonal variations in raw water quality. Further investigation is required in this area.

Having said that, there is little doubt that nanofiltration is a viable method for the effective control of THM concentrations in these small coloured water supplies.

ACKNOWLEDGEMENT

The author wishes to thank Mr J Cockburn, Managing Director-Water Services, North of Scotland Water Authority for granting permission for this paper to be presented.

CHLORAMINATION: THE BACKGROUND TO ITS INTRODUCTION, AND THE MANAGEMENT OF NITRITE FORMATION IN A WATER DISTRIBUTION SYSTEM

J. A. Haley

Water Quality Manager
The York Waterworks Plc
Landing Lane, York, YO26 4RH

1   INTRODUCTION

York Waterworks Plc supplies potable water to a population of approximately 177,000 in the City of York, surrounding commuter villages and rural districts. Water is abstracted from the River Ouse and treated either by coagulation/clarification, rapid filtration and slow sand filtration (70%), or coagulation/upflow clarification and rapid filtration (30%). The slow sand filters have recently been converted to incorporate a layer of granular activated carbon (GAC), the so-called Thames Sandwich.

The objective of the trial was to determine whether chloramination could enable the Company to establish a significant chlorine residual (>0.1 mg/l) throughout its distribution network in order to address other water quality issues.

1.1   Previous Chlorination Practice

After the completion of all physical treatment processes water was disinfected by free chlorine (hypochlorous acid). Sufficient chlorine was injected to achieve a free chlorine concentration of 1.0- 1.5 mg/l within the contact tank. After 1.5hrs contact excess chlorine was removed by sulphur dioxide to leave a residual free chlorine concentration of 0.1 mg/l; this was associated with approximately 0.1 mg/l combined chlorine due to reaction of chlorine with natural ammonia present in the treated water. (combined chlorine is more persistent within distribution networks) The free chlorine residual gave rise to a number of customer complaints from properties near to the major trunk mains; however, the residual did not persist significantly within the distribution network.

Work associated with the installation of GAC in the slow sand filters, for pesticide removal, necessitated a change to the point of application of chlorine from before, to after a bank of rapid gravity filters (RGFs) for 30% of the works output. Biological activity that developed within those RGFs was found to effectively remove the natural ammonia, leading to a loss of the combined residual chlorine. As a consequence, in the summer months it was necessary to supplement the chlorine residual by applying a small dose of chlorine dioxide to maintain satisfactory microbiological water quality.

1.2   Microbiological Factors

With the loss of combined chlorine residual, significantly higher total viable counts (TVC) (22°C) counts were encountered when water temperatures rose during the summer months. In a number of cases this was associated with taste and odour complaints from customers. Investigations using R2A agar – 7day counts, indicated increased biofilm activity within the distribution system. There appeared to be an increasing trend in both the number of total coliform detections from the distribution system, and the number of presumptive coliform detections requiring follow-up investigations. (Figure 1)

1.3   Options Considered

1.3.1   Increased use of chlorine dioxide. The increased use of chlorine dioxide, both in terms of the period of use and doses applied was considered. This option was not proceeded with on the basis of its effect on costs of production, more difficult control and monitoring and concerns about chlorate as a by-product.

1.3.2   Chloramination – using monochloramine. The main concerns related to the potential to develop objectionable chlorinous taste and odour if ammonia dosing was not optimised, and also the risk of generating significant concentrations of nitrite as a result of bacterial oxidation of ammonia liberated by monochloramine. The sensitivity to chlorine of customers who lived towards the periphery of the distribution system, where a significant chlorine residual is unlikely to have existed, was a concern. These customers represented a large potential number of complaints, triggered as the increased chlorine residual reached them.

Figure 1   *No. of total coliform failures/presumptive coliform investigations*

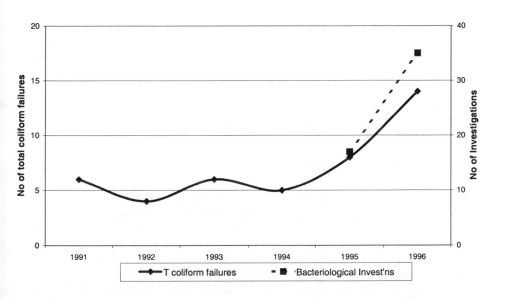

1.4   Nitrite

Nitrite formation was a major concern in the consideration of the options, principally because the Company did not wish to exchange coliform detections for infringements of the nitrite Prescribed Concentration or Value (PCV).

1.4.1   The UK (& EU) Water Quality Standard. The PCV for nitrite in the UK is 0.1 mg/l, as derived from the EU Drinking Water Directive. This standard is to be revised by the Revised Drinking Water Directive, expected to be ratified in late 1998. The revised standard will be based on a formula linking the concentrations of nitrite and nitrate as currently defined by WHO.

1.4.2   The World Health Organisation Standard. The WHO standard for nitrite is expressed by the following formula:

$$\frac{\text{mg/l nitrite}}{3} + \frac{\text{mg/l nitrate}}{50} \leq 1$$

The above standard is derived from the relative potency for methaemoglobin formation by nitrite and nitrate. [1]

2   PRELIMINARY INVESTIGATIONS

The objectives of the investigations were to determine a suitable point of injection for ammonia, and define the operational parameters involved in generating chloramines. Further studies were undertaken to determine the potential increase in survival time of chloramine residuals when compared to free chlorine, to assess the effects on taste and odour of treated water, and if possible quantify the potential for nitrite formation.

2.1   Operational Considerations

Operationally the most appropriate point of application of ammonia was considered to be at the outlet to the main contact tank. Investigations showed there to be sufficient free chlorine available at that point to be able to generate 0.8mg/l combined chlorine. This point provided the best process option with an existing flashmixer, dosing line, and monitoring system.

After laboratory investigation the stoichiometric "N" requirement was confirmed as 0.2 x available free chlorine. [2]

2.2   Laboratory Investigations

Laboratory trials were undertaken to assess the stability of chlorine species formed by the addition of ammonia, the taste and odour of treated water and the formation of nitrite. The trials were undertaken using 1 litre samples in acclimatised amber glass bottles. To assess the effects of water temperature trials were run in parallel at 3°C and 22°C for a total of 7 days. This period was chosen to reflect the estimated maximum retention time for water in the distribution network.

For the first six hours residual chlorine determinations were undertaken at two-hour intervals. Subsequently analyses were undertaken on a daily basis, including a qualitative assessment of taste and odour. Figure 2 summarises one trial run at 22°C, which was expected to be the worst case with respect to residual chlorine decay and nitrite formation. Figure 3 summarises one trial run at 3°C, which was expected to return acceptable results for chlorine residual decay and nitrite formation.

2.3 Preliminary Consultations

The next phase of the trial was expected to involve the full-scale switch from free chlorine to monochloramine residual in the water leaving the treatment works. With the potential to generate customer complaints during this procedure and the medium term possibility of nitrite formation, a program of consultation was undertaken. Initially this involved the Drinking Water Inspectorate, the Environmental Health Officers of City of York and Selby District Councils and the Consultant in Public Health Medicine at North Yorkshire Health Authority.

At this stage the project was more widely disseminated within the Company, with specific advice developed for those employees who would be expected to deal with customer enquiries and complaints.

Figure 2   *Laboratory chloramination trials at 22°C*

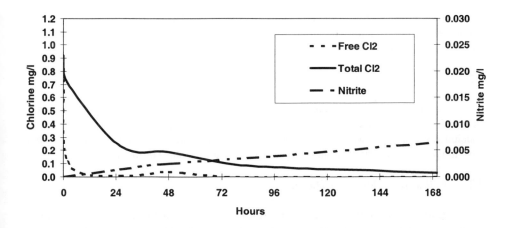

3   IMPLEMENTATION OF CHLORAMINATION

The initial phase of the trial was to be conducted using ammonium chloride solution as the ammonia source. Use of ammonia gas was excluded on safety grounds and the complexity of the equipment required to eliminate problems due to precipitation of hardness in the dosing lines. Monitors sensitive to chloramines were installed and commissioned. Mixing and stock tanks and dosing pumps for ammonium chloride solution were also installed. Dosing pump speed was made proportional to treated water flow. Pump stroke was to be manually controlled, as this variable would be related to the free chlorine available, a parameter determined by an existing control system.

Figure 3   *Laboratory chloramination trials at 3°C*

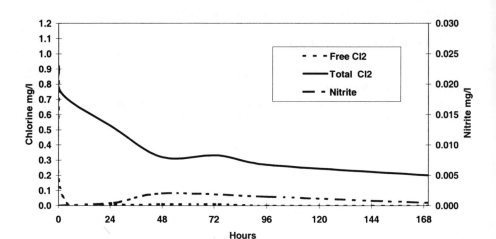

Sample points within the distribution system were selected as being suitable for monitoring throughout the likely extent of the trial. These were located at either employees' homes or public buildings where entry could be guaranteed for the duration of the trial

3.1   Preliminary Dosing Checks

Short timescale trials were run to fully test the dosing and monitoring systems and ensure that the automatic systems operated as they were expected to. These trials were concluded successfully without identifying any major potential operational problems. This gave the project team the confidence to proceed to the final stage of the trial.

3.2   Full-scale Chloramine Dosing

Ammonia dosing commenced in April 1997 with an initial target of 0.4 mg/l monochloramine leaving the treatment works. Over subsequent days this was gradually increased to 0.55 mg/l. Monitoring of the progress of the chloramine residual through the distribution network commenced, including analysis for nitrite.

3.3   Operational Considerations over the First Year of the Trial

Once the initial changeover had been made, the ex-works chloramine residual was adjusted in response to water temperature. Experience had demonstrated this to be the best means of maintaining stable residuals within the trunk main network and of avoiding customer complaints. The monitoring of chlorine residuals continued and where customer complaints occurred these were dealt with on a case by case basis.

Figure 4   *Example of trend in chlorine residual in water tower, remote from WTW*

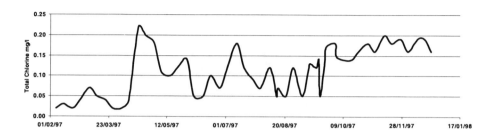

4   REVIEW OF RESULTS FROM FULL-SCALE TRIAL

The results of the water quality monitoring undertaken by the Company are reviewed, together with an assessment of the impact on customers and other distribution system issues.

4.1   Effect on Chlorine Residuals in the Distribution Network

The effect of the change in residual chlorine species on the trunk mains system was

rapid and significant (Figure 4). The response of residuals in water towers and reservoirs lagged a little behind but steady progress was made over the first few weeks. Results from distribution mains were much more variable reflecting their diverse materials and condition (Figure 5).

4.2   Customer Perception

Initial feedback from customers was positive. Where we had had long-standing complaints from customers close to the major trunk mains the comments were entirely positive with a perceived reduction in chlorinous taste and odour.

As the trial progressed there was a gradual increase in customer enquiries from areas of the distribution system where it is unlikely significant chlorine residuals had ever existed. From our investigations it was clear that some of these individuals were very sensitive to chlorine species even if the total chlorine residual was only around 0.05 mg/l.

In a few areas complaints of very pronounced earth/musty odours were received. These were confined to dead-end or looped mains with relatively low rates of flow. Water quality investigations showed this to be due to bacteria, rather than yeasts or moulds, and was associated with total loss of the chlorine residual and complete conversion of ammonium, through nitrite, to nitrate. The biofilm was not effectively removed by flushing, only air-scouring or swabbing, with a chlorinated sponge, was successful. In a few cases it was necessary to CO_2 scour the service pipes to affected properties.

4.3   Nitrite Formation

Nitrite formation has not been a significant problem in the distribution system as a

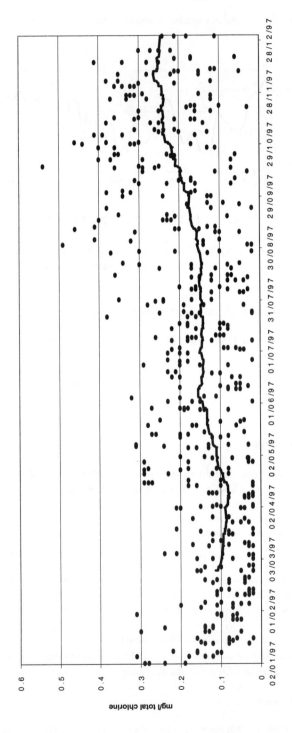

Figure 5   *Trend in chlorine residuals within the distribution system*

Figure 6 *Example of downstream series sampling monitoring nitrite production*

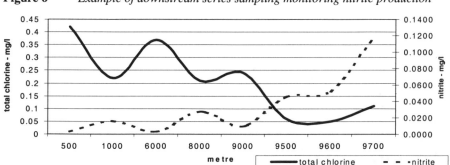

whole. However, where significant chlorine residuals have not been established in highly reticulated estates and in some rural villages, nitrite concentrations in excess of the PCV have been observed. Typically water arrives at the estate or village in good condition, with low concentrations of nitrite and a significant chlorine residual. However, the residual is rapidly dissipated, and nitrite concentrations rise rapidly, often within a few hundred metres. (see Figure 6) In most cases air-scouring of the affected main has been found to be the most effective remedial action in restoring good water quality.

From the data gathered so far, it appears that significant chloramine residuals inhibit the formation of nitrite. Figure 7 demonstrates this for the York distribution system. This supports the findings obtained by Thames Water in their work with chloramine in the London distribution system.[2] The critical chlorine concentration appears to be between 0.15-0.20 mg/l of monochloramine.

No nitrite concentrations of health significance have been generated, and under the propsed Revised Drinking Water Directive no PCV infringements would have occurred. Monitoring is continuing as before, but is now supplemented by means of R2A agar counts and the opportunistic analysis of mains cut-out samples which are swabbed to recover any adherent biofilm.

5 OVERALL CONCLUSIONS

After a year of the trials it is too early to draw final conclusions. However very significant progress has been made in some of the target areas.

5.1 Microbiological Activity

As one of the key drivers for the trial, a positive improvement in bacteriological quality was anticipated. Figure 8 demonstrates the very significant progress which has been made in this area. Only three samples were found to contain total coliforms during the year. Of these one was taken before the trial, and the other two were related to a biofilm problem in a rarely used pumping station. There has been a corresponding

Figure 7   *Nitrite production: - comparison with chlorine residual*

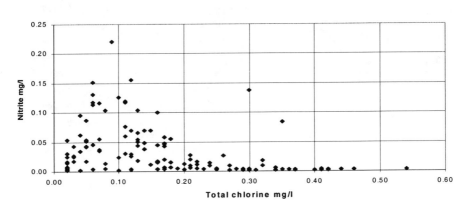

Figure 8   *Effects of adoption of chloramination on bacteriological parameters*

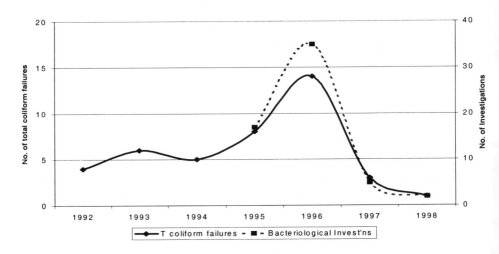

decrease in the number of TVCs isolated from customer tap samples. Operationally the need to respond to significantly lower numbers of "false" coliform detections has been of benefit

5.2   Customer Reaction

A significant number of customers, in particular those close to trunk mains, are more content with the chloramine residual as it is less pungent than the lower concentrations of free chlorine they previously experienced. However, the gradual increase of residuals in peripheral areas of the distribution system continues to trigger complaints, although most customers accept the Company's reasons for taking the action it has.

5.3   Other Benefits

5.3.1   Asellus. There is evidence that the prevalence of this freshwater organism is decreasing. Surveys undertaken in early 1998 showed the organism to be confined to a few pockets in dead-end mains, even within the small areas where their presence had been widespread when survey in 1995. No complaints have been received about the organism for the past year during which period several would have been expected.

5.3.2   THMs. The concentrations of THMs in samples taken from the distribution system are significantly lower than those previously experienced. Without doubt some of this effect will be due to the reduction in organic carbon in treated water due to the commissioning of additional GAC capacity, but some additional reduction appears to have occurred due to the change in residual chlorine species.

5.4   Overall Appraisal

Overall the trial has appears to have been a success, achieving the goal of reduced microbiological activity with a limited impact in terms of compliance with the nitrite standard. Fundamental to the minimisation of nitrite concentrations is careful control of the ammonia to chlorine ratio. It is apparent that the achievement of target residuals throughout the distribution network is taking longer than projected, but steady progress is being made.

Acknowledgements

The author recognises the assistance of many other people critical to the success of this project. Particular reference is made to the contribution of Tim Claydon, Water Quality Scientist, who was responsible for much of the day-to-day operation of the trials and whose analysis and collation of the data was invaluable. The assistance of members of the Company's Water Quality, Distribution & Production Departments is also recognised.

References

1.   World Health Organisation, Geneva; Guidelines for drinking water quality, 2nd ed. vol. 1 – Recommendations. 1993. ISBN 92 4 154460 0, 53.

2.   Holt, D., Todd, R.D.,Delanou, A., & Colbourne, J.S.; A study of nitrite formation and control in chloraminated distribution systems. Proc. of AWWA Water Quality Technology Conference 1995; pp 1427-1439.

Advances in Analysis and Monitoring

TRACE-LEVEL DETERMINATION OF BROMATE IN DRINKING WATER BY IC/ICP-MS

F. Sacher, B. Raue and H.-J. Brauch

DVGW-Technologiezentrum Wasser
Karlsruher Strasse 84
D-76139 Karlsruhe
Germany

1 INTRODUCTION

Bromate is formed during ozonation of bromide-containing waters.[1-5] As bromate is a suspected carcinogen, the new European Drinking Water Directive will include a parametric value for bromate which will probably be 10 μg/l. Within the research project "Laboratory and field methods for determination of bromate in drinking water" supported by the European Commission (contract no. SMT4-CT96-2134) different analytical methods are being developed, improved and validated in order to ensure the reliable determination of bromate in drinking waters at the low μg/l level. Partners in this project are Anjou Recherche (coordinator, France), CIRSEE (France), University of Poitiers (France), KIWA N.V. (The Netherlands), SVW (Belgium), University of Sheffield (UK), Yorkshire Environmental Ltd. (UK) and DVGW-TZW (Germany). In this paper the development and optimisation of the coupling of ion chromatography (IC) and inductively-coupled plasma mass spectrometry (ICP-MS) for the trace-level determination of bromate in drinking water is presented.

2 DEVELOPMENT OF AN IC/ICP-MS METHOD

2.1 Optimisation of the ICP-MS System

Prior to the determination of bromate by coupling an ion chromatographic system to an ICP-MS, the parameters of the ICP-MS have to be optimised in order to obtain maximum sensitivity. As a rule, this optimisation procedure is done on a multi-element level and is not focussed on one specific element. In Table 1 the optimized ICP-MS parameters are summarized.

Table 1 *ICP-MS parameters for the determination of bromine species*

ICP-MS:	Perkin Elmer Elan 6000
RF power:	1150 W
plasma gas:	argon
plasma gas flow:	15 l/min
auxiliary gas flow:	1 l/min
nebulizer gas flow:	0.875 l/min (cross-flow)
flow rate:	1 ml/min
detector:	pulse counting mode
lens voltage:	auto lens

2.2 Determination of Bromine Species by ICP-MS

Two isotopes of bromine are found naturally: ^{79}Br with an exact mass of 78.918 amu and an abundance sensitivity of 50.69% and ^{81}Br with a mass of 80.916 amu and an abundance sensitivity of 49.31%. As the abundance of both, ^{79}Br and ^{81}Br is nearly 50%, the sensitivity of ICP-MS which measures at one time only the response of one selected isotope always is reduced. Another problem for a sensitive determination of bromine arises from the first ionisation potential of bromine of 11.814 V, which is a very high value compared to most of the other elements of the periodic table, resulting in an ionisation yield for bromine of only 5% under common ICP conditions.

Interferences on the masses of the two bromine isotopes may be derived from a ^{38}Ar^{40}ArH cluster for ^{79}Br and a ^{40}Ar^{40}ArH cluster for ^{81}Br. As the natural abundance of ^{38}Ar is only 0.06% of the natural abundance of ^{40}Ar, the background noise for the mass number 79 is much lower than for the mass number 81. This theoretical prediction is easily confirmed by measuring the background noise on both mass numbers, ^{79}Br and ^{81}Br. Therefore, sensitive detection of bromine species by ICP-MS is preferably done via the ^{79}Br isotope.

In Table 2 the recoveries of bromine in bromide and bromate with the ICP-MS are given for two different matrices, 1% nitric acid and the effluent of a suppressor unit, which was fed with a 8 mM bicarbonate/carbonate eluent. All experiments were performed at a concentration level of 100 μg/l Br. As reference standard an IC standard with 1 g/l Br as NaBr in pure water was used. A suppressor unit is used in ion chromatography for the suppression of the conductivity of the eluent. Within this unit cations are exchanged for H$^+$ ions which recombine with the anions of the eluent, e.g. CO_3^{2-} or HCO_3^- resulting in a lower conductivity of the eluent and an increase in sensitivity using conductivity detection. As a second effect, this cation exchange step also decreases the salt load of the eluent before entering the ICP-MS, resulting in a higher stability of the nebulizer system. Therefore, the use of a suppressor unit is also recommended when using ICP-MS detection. The 8 mM bicarbonate/carbonate eluent which was used to feed the suppressor unit is a common eluent used for the ion chromatographic separation of bromate.

It can be seen from Table 2 that the recovery of bromine is higher for bromate than for bromide. This might be a result of the different oxidation state of bromine in the two species which is +V for bromate and -I for bromide. Due to the fact that the bromine content of bromate is only about 62%, the mass sensitivity of course is higher for bromide than for bromate.

In Table 3 the limits of detection, identification and determination as well as the repeatability for the analysis of bromide and bromate using the ICP-MS are given. Calculation of these parameters was done according to the German standard procedure DIN 32645, using a calibration with ten concentration levels in the range of 1 to 10 μg/l. For the determination of the repeatability a tenfold injection of one sample with 100 μg/l Br was done within one day. Again, all measurements were performed in the effluent of a suppressor which was fed with a 8 mM carbonate eluent.

Table 2 *Recoveries of bromine in different bromine species and different matrices*

matrix	species	recovery in %
1% HNO$_3$ (pH < 1)	IC standard	101
	KBr	102
	KBrO$_3$	120
suppressor's effluent (pH ≈ 4)	IC standard	101
	KBr	100
	KBrO$_3$	126

Table 3 *Limits of detection, identification and determination and repeatability for the analysis of bromide and bromate using the ICP-MS system (for details see text)*

species	limit of detection in µg/l	limit of identification in µg/l	limit of determination in µg/l	repeatability (c=100 µg/l Br, N=10)
bromide	0.23	0.46	0.76	0.78%
bromate	0.11	0.23	0.39	0.77%

The data prove that both, the sensitivity and the repeatability of the ICP-MS system are excellent. The correlation coefficients show that the detection is linear over the working range under investigation. Supplementary measurements prove that the linearity holds true over a concentration range of more than three decades. Hence, the preconditions for a sensitive and reliable determination of bromate by IC/ICP-MS are fulfilled.

2.3 Coupling of the ICP-MS to an Ion Chromatographic System

The scheme of the whole experimental set-up for the determination of bromate by IC/ICP-MS is shown in Figure 1. The eluent is pumped by a LC pump through the injection system, which is a Rheodyne 6-port valve with a 100 µl sample loop, and subsequently on a precolumn and a separation column, where the ion chromatographic separation of the analytes takes place. Then, the eluent passes the suppressor unit which is regenerated by sulfuric acid. The effluent of the suppressor unit is directly coupled to the cross-flow nebulizer system of the ICP-MS. Details of the IC conditions are summed up in Table 4. Data acquisition was done by the Perkin Elmer ELAN software with the parameters which are also given in Table 4. However, for data processing the TurboChrom software from Perkin Elmer was used, which allows a better and more convenient evaluation of the data. Hence, prior to data evaluation the data sets acquired by the ELAN software were transformed into another, more suitable format. This can be done by some additional software which was also provided by Perkin Elmer.

Figure 1 *Experimental set-up for the IC/ICP-MS coupling*

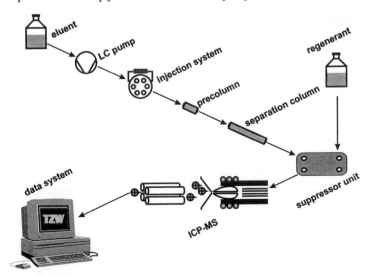

Table 4   *IC conditions and parameters for data acquisition for the determination of bromide and bromate by IC/ICP-MS*

pump:	Perkin Elmer 200LC Bio pump
precolumn:	Dionex IonPac AG9A-SC
separation column:	Dionex IonPac AS9A-SC
eluent:	8 mM Na_2CO_3/$NaHCO_3$
flow rate:	1 ml/min
sample volume:	100 μl
suppressor:	Dionex ASRS-I
regenerant:	25 mM H_2SO_4
regenerant flow:	5.6 ml/min
data acquisition:	ELAN software
scan mode:	peak hopping
measurement unit:	counts
sweeps/reading:	1
readings/replicate:	120
replicates:	1
dwell time:	2500 ms
integration time:	300000 ms

Figure 2 presents a chromatogram of a standard solution with 10 μg/l bromate and bromide in deionized water. The figure illustrates that with the Turbochrom software and the transformed data sets peak labelling, manual and automatic baseline setting, peak integration as well as generation of calibration tables and automatic report generation can easily be done.

2.4 Performance Data

For the calculation of the performance data of the IC/ICP-MS method, a calibration was carried out in spiked deionized water with ten concentration levels in the range of 1 to 10 μg/l bromate and 10 to 100 μg/l bromide, respectively.

Figure 2 *Chromatogram of a standard solution with 10 μg/l bromate (A) and bromide (B)*

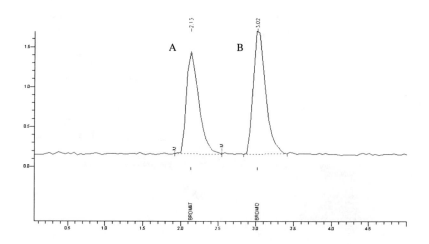

Table 5 *Limits of detection, identification and determination and repeatability for the analysis of bromide and bromate using IC-ICP-MS (for details see text)*

species	limit of detection in μg/l	limit of identification in μg/l	limit of determination in μg/l	repeatability
bromide	1.1	2.3	4.2	7.4%
bromate	0.28	0.57	1.0	9.8%

In Table 5 the limits of detection, identification and determination are given. Calculation of these parameters was done according to the German standard method DIN 32645. As it can be seen, detection limits for bromate of about 0.3 μg/l can be achieved without preconcentration. In Table 5 also data for the repeatability of the method in deionized water are presented. For the determination of the repeatability a tenfold injection of one sample with 1 μg/l bromate and 10 μg/l bromate was done within a time period of one day. Again, the data prove that the repeatability of the IC-ICP-MS method is excellent even at low concentration levels.

3 IC/ICP-MS WITH ON-LINE PRECONCENTRATION

Combining the IC/ICP-MS method with an on-line preconcentration step[6,7] provides a very powerful tool for the sensitive detection of bromate. Using the on-line preconcentration technique for the analysis of real water samples chloride and sulphate have to be removed prior to the preconcentration step. This is done by using barium and silver cartridges. Subsequently, a sample volume of 2 ml was preconcentrated on a Dionex AG9-HC guard column and analysed with the method described before.

Figure 3 presents a chromatogram of a drinking water which was spiked with 30 ng/l of bromate and analysed with the on-line IC/ICP-MS technique. It can be seen that even this low bromate concentration could be traced without difficulties.

Calculating the limit of detection according to the German standard method DIN 32645 for bromate 18 ng/l were attained for the on-line IC/ICP-MS method. Nevertheless, for routine analysis this on-line method is exaggerated and for the control of the parametric value of the new European drinking water directive, the IC/ICP-MS method without preconcentration step is sufficient.

4 COMPARISON OF THE IC/ICP-MS AND THE IC/CD METHOD

In Figure 4 the results of the IC/ICP-MS method for different ozonated waters and drinking waters are plotted versus the respective results of an ion chromatographic method using conductivity detection. For the IC/CD method samples were tenfold preconcentrated using an off-line method which includes a time-consuming evaporation of the water.[5] Analysis with the IC/ICP-MS method was done without any sample pretreatment, whereby the time demand per sample was about five minutes.

It can be seen from Figure 4 that the correspondence of data is very good for the whole concentration range between 1 and 20 μg/l. This result again stresses the excellent performance of the IC/ICP-MS method.

Figure 3   *Chromatogram of a standard solution with 30 ng/l bromate (A) in drinking water analysed with the on-line-IC/ICP-MS technique*

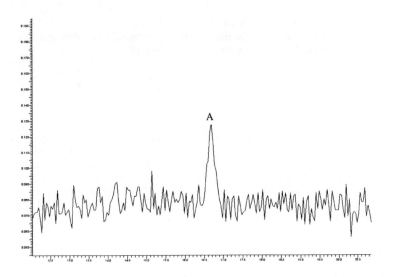

Figure 4   *Bromate concentrations measured with IC/ICP-MS vs. bromate concentrations measured with IC/CD*

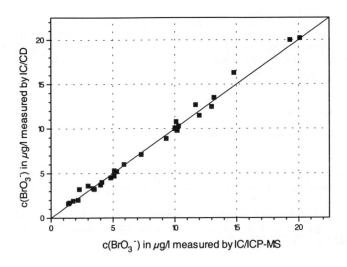

References

1. W. R. Haag and J. Hoigné, *Environ. Sci. Technol.*, 1983, **17**, 261.
2. J. C. Kruithof and J. C. Schippers, *Water Supply*, 1992, **11**, 121.
3. S. W. Krasner, W. H. Glaze, H. S. Weinberg, P. A. Daniel and I. N. Najm, *J. Am.Water Works Assoc.*, 1993, **85**, 73.
4. U. v. Gunten and J. Hoigné, *Environ. Sci. Technol.*, 1994, **28**, 1234.
5. F. Sacher, A.Matschi and H.-J. Brauch, *Acta hydrochim. hydrobiol.*, 1995, **23**, 26.
6. R. J. Joyce and H. S. Dhillon, *J Chromatogr. A*, 1994, **671**, 165.
7. C. Bruggink, W. J. M. v. Rossum and J. G. M. M. Smeenk, *H_2O*, 1995, **28**, 343.

Acknowledgements

The financial support by the European Community (contract no. SMT4-CT96-2134) is gratefully acknowledged.

FIELD AND LABORATORY METHODS FOR BROMATE DETERMINATION IN DRINKING WATER

V. Ingrand, M. Aubay and M.C. Müller

Central Laboratory
Anjou Recherche
1, Place de Turenne
94417 Saint Maurice

1 INTRODUCTION

In order to meet increasingly strict quality criteria, drinking water producers need to adapt treatment stages to remove organic, mineral and biological compounds. To achieve this, they have, among other means available, a wide range of oxidants and disinfectants which are applicable according to given matrices. Consequently, ozone, a powerful oxidant, is widely used in Europe. In drinking water treatments, it provides disinfection (it is particularly efficient for the deactivation of the *Giardia* cysts), the removal of organic micropollutants, iron, manganese and the improvement of organoleptic qualities.[1] In addition, the use of ozone prevents the formation of chlorinated derivatives.

However, in raw waters containing bromides, ozone can, under certain conditions, lead to the formation of by-products, such as bromates. These latter can also result from disinfection using sodium hypochlorite. Following toxicological studies, CIRC classified bromates in group 2B, being a substance that is potentially carcinogenic for humans. This risk led to the establishment of a provisional guide value fixed at 25 μg/l in drinking water.[2] This Maximum Permissible Concentration will be fixed at 10μg/l in the future European directive.

Consequently, it has become urgent to develop new, more sensitive and more reliable analytical methods to monitor the ozonation and control contents in the drinking waters. To this end, the European Community has financed a project named "Laboratory and field methods for determination of bromate in drinking water". This project is co-ordinated by Anjou Recherche's Central Laboratory.

2 PRESENTATION OF THE EUROPEAN PROJECT

Few laboratories currently have reliable, robust and applicable routine methods able to quantify bromates at 2.5 μg/l or less.

Apart from Anjou Recherche, this European project implicates 7 other water

producer or water research centre partners (French, Belgian, German, English and Dutch). The goals of the European project are as follows :
- Improve the laboratory method based on ionic chromatography coupled with conductimetric detection. The developed method must be robust, user friendly and less sensitive to interfering agents so as to be able to quantify 1 μg/l of bromates.
- Develop an alternative method able to quantify 1 μg/l of bromates in raw and drinking waters and confirm the presence of bromates in raw waters.
- Develop a robust and low-cost field method permitting operational monitoring.
- Study the stability of bromates in different matrices.
- Validate the developed methods.

Following this project, a reference method shall be proposed to the CEN TC 230 standardisation committee which is responsible for standards concerning this element.

3 BROMATE ANALYSIS

Scientific literature provides different methods to analyse bromates in matrices at a wide range of concentrations. Until now, few of the described colorimetric or fluorimetric methods made it possible to reach a quantification limit adapted for use in drinking waters. The use of different coloured indicators led to quantification from 5 to 0.13 mg/l.[3,4,5] But these reactions are not very sensitive and subject to interfering agents. Gordon *et al*[6] attained a detection limit of 0.7 μg/l through the oxidation of a coloured indicator : chlorpromazine. Detection by fluorescence quenching with other coloured indicators also seem sensitive as Guoquan et al have quantified 2μg/l of bromate in potassium chlorate.[7]

Coupled with a fractionating/concentration stage, impulsion polarography presents a 2 μg/l detection limit.[8] Amperometry permits approximately 130 μg/l to be dosed.[9] The method most generally used to analyse bromates in waters is ionic chromatography coupled with conductimetric detection. This technique can be used either with or without pre-concentration. The quantification limits attained are respectively 1-2 μg/l[10,11] and 10 μg/l.[12] Coupling ionic chromatography and a detection method such as ICP-MS leads to the detection limit being reduced to concentrations of around 0.1 - 0.2 μg/l.[13,14] However, these complicated and expensive methods cannot be used for the routine laboratory determination of bromates. This implied the need to develop alternative, simpler methods. This study included the two following stages :
- Development of a field colorimetric method adapted to the continuous flow method developed by Gordon *et al.*[6]
- Development of an alternative laboratory method based on the separation of bromates by ionic chromatography followed by a post-column reaction between the bromates and a coloured indicator. The reaction product is measured using a UV/Visible detector.

4 EQUIPMENT AND METHOD

4.1 Reagents

The standard solutions are prepared using quality reagents for analysis ($KBrO_3$ and

H$_3$BO$_3$, Merck; HCl 36%, Prolabo; NaOH 50%, J.T. Baker; Chlorpromazine, Aldrich). Ultra-pure water was obtained using an ion exchanger cartridge demineralisation system.

4.2 Equipment

4.2.1 Field Method. The analyses were carried out using a UVIKON 941 spectrophotometer (Kontron) in a glass cell with a 5 cm optical path length. The work wavelength was 525 nm.

4.2.2 Laboratory Method. The analyses were carried out using a DX500 chromatograph (Dionex) equipped with a conductimetric detector, an anion self-regenerating supressor and a UV/Visible detector. For the post-column reaction, a mixing tee is placed on the conductimeter outlet. A pressurised tank introduces the coloured indicator into the mixing tee. The reaction between the bromates and the coloured indicator (chlorpromazine) takes place in a 375 μl reaction loop (Dionex). The chlorpromazine is introduced in solution form at 0.6 g/l in the presence of HCl 3 mol/l.

AG9SC type columns are used for the pre-concentration column and the guard column and a AS9SC for the analysis column. A low ionic force eluant (H$_3$B0$_0$ 40 mM/NaOH 20 mM) is used for the analysis. A high ionic force eluant (H$_3$B0$_3$ 250 mM/NaOH 100 mM) permits the separation of the most retained anions through the use of a gradient step. These two eluants are filtered on 0.45 μm and vacuum degassed prior to use.

The samples are filtered on a silver cartridge (Dionex), then an H$^+$ resin (Alltech) and finally on a 0.45 μm filter (Millipore).

5 RESULTS

5.1 Development of the Field Method

5.1.1 Principle of the Spectrophotometric Method. Chlorpromazine is an indicator from the phenothiazine family that has a nitrogen atom and a sulphur atom susceptible to oxidisation. In an acid environment, the chlorpromazine is oxidised by the bromates into a coloured compound. The intensity of the colour is measured by spectrophotometry.

5.1.2 Optimisation of Operating Conditions. The reagents must be added in the following order : bromate (water to be analysed) - chlorpromazine - HCl. If the bromates are mixed with acid prior to adding the chlorpromazine, the obtained absorbance is approximately 75% lower than that obtained in the pre-defined order. No waiting time between each reagent is needed. The optimal concentrations are 200 mg/l of chlorpromazine and 1 mol/l of hydrochloric acid. The study of the reaction kinetic demonstrates that the product formed is stable for at least 20 minutes in ultra-pure water and in real water.

In these conditions, it is possible to obtain the calibration curves presented in figures 1 and 2. For low calibration (figure 1), points 0.1 and 2 μg/l correspond to an average of 3 measurements. The bars associated with these points represent the absorbance variation observed during these three measurements. Figure 1 shows that the quantification of bromates is not possible for concentrations below 10 μg/l as the analysis is not sufficiently repeatable. There is a good linearity for greater concentrations.

In order to improve the limit of quantification of the method, the acidity conditions were studied by changing the nature of the acid : sulphuric acid, phosphoric acid and nitric acid. These tests did not improve the analysis. Consequently, the initial conditions were retained.

5.1.3   *Study Using Real Waters.* Spiking using bromates at 0, 5, 10 and 25 μg/l was carried out on different real waters with known mineralization and which did not contain bromates (table 1). Table 1 shows that the quantification of bromates at concentrations lower than 10 μ/l is not possible. In the absence of bromates (blank test), the absorbance obtained corresponds to a bromate equivalence concentration of between 2 and 5 μg/l. In tap water and Evian water, spiked with 5 μg/l, the concentration obtained corresponds well to the sum of the blank and the spiking. This doping was not found in Contrexeville water. Additions of 10 and 25 μg/l are found in all the waters. These results confirm that it is not possible to quantify bromates at concentrations lower than 10 μg/l in real waters. The results obtained at concentrations lower than 10 μg/l could be due to the limit of quantification above this value or to the presence of interfering agents.

Figure 1 *Low BrO$_3^-$ Calibration Curve*

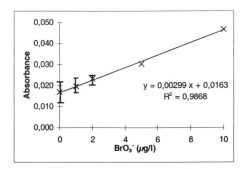

Figure 2 *High BrO$_3^-$ Calibration Curve*

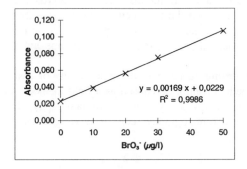

The Evian water was then spiked with 30 μg BrO_3^-/l and analysed 10 times. The average is 28.4 μg/l with a 3.5% variation coefficient. The analysis is repeatable but the divergence between the theoretical value and the observed value is 13.6%. This divergence is probably due to elements interfering in the analysis.

Table 1 *Bromate Determination in Different Matrices*

Bromate spiking (μg/l)	Tap water (low mineralisation)	Evian water (medium mineralisation)	Contrexeville water (high mineralisation)
0	2	3	5
5	7	9	6
10	11	12	10
25	23	24	23

5.1.4   Study of interfering agents. Tests showed that the HCO_3^-, NO_3^-, SO_4^{2-}, Cl^-, HPO_4^{2-}, Br^- anions and silica do not interfere in the method at concentrations normally found in drinking waters. Bibliography indicates that this method is interfered by metals, nitrites and chlorites and gives the means for their removal.[6] Sulfamic acid transforms nitrites into nitrates, but there is no treatment to totally remove chlorites. Chlorates begin to interfere with the analysis at concentrations of 1,000 μg/l, but this content is rarely encountered in drinking water. Hypochlorite ions also interfere in the analysis. Different compounds known for their action with regards hypochlorites were tested. Oxalic acid and malonic acid, either taken alone or coupled with sodium cyclamate, present a satisfactory efficiency in doped ultra-pure water but not in real water. Thioacetamide, a powerful reducing agent, overcomes the ClO^- interference but also partially reduces the bromates to the concentrations needed to reduce the hypochlorites. The association of thioacetamide and glycine or sodium cyclamate does not improve the results.

Certain compounds partially or completely overcome one or more interfering agents but none completely remove all of them. The aim is not use a product per type of interfering agent as this process would be too complicated to organise and would lead to the sample undergoing a high level of dilution. As the removal of hypochlorites and other disinfection products is not possible, the method has been applied to 5 ozonated waters sampled upstream from the disinfection stage (table 2). The concentration divergences obtained between the two methods range from 22 to 380% without treatment and 19 to 200% after the removal of metals by passage over H^+ resin (Alltech). This treatment is therefore not sufficient to overcome all the interferences.

Given the large number of interfering agents and the impossibility of easily removing them, the chlorpromazine method cannot be chosen as a field method. As the interference of elements can be removed by ionic chromatography, the chlorpromazine method, coupled with chromatography, can be envisaged as an alternative laboratory method.

Table 2 *Bromate Determination in 5 Ozonated Waters*

Origin of waters	Bromates Concentration ($\mu g/l$)				
	Ionic Chromatog. (I.C.)	Direct Analysis		Analysis after treatment	
		BrO_3^-	*Accuracy / I.C. (%)*	BrO_3^-	*Accuracy / I.C. (%)*
Water n°1	27	33	*22*	32	*19*
Water n°2	8	14	*75*	15	*88*
Water n°3	19	91	*380*	53	*180*
Water n°4	6	20	*233*	18	*200*
Water n°5	7	13	*86*	13	*86*

5.2 Development of an Alternative Laboratory Method

The development of an alternative method based on ionic chromatography coupled with a post-column chemical reaction associates the advantages of the two techniques: the separation of elements by chromatography and the specificity of the colorimetric reaction with the chlorpromazine.

5.2.1 Development. The analysis using the spectrophotometric method is carried out by the successive addition of chlorpromazine and then hydrochloric acid. For the post-column reaction, the addition of a single reagent is preferable. A first series of batch tests demonstrated that the reaction with the bromates develops in an identical manner if the chlorpromazine (CLP) and the acid (HCl) are added one after the other or simultaneously. However, the relation of concentrations between these two compounds influences the stability of the mix. The greater the acid and chlorpromazine concentrations, the lower the stability of the mix. This is translated by the increase in the baseline over time. Given this phenomenon and the technical limitations (tank volume that can be pressurised and applicable pressure), the optimal operating conditions are achieved by a chlorpromazine 0.6 g/l / HCl 3 mol/l mix. In these conditions, the direct injection of a 5 $\mu g/l$ reference solution using a 200 μl loop provides the chromatogram shown in figure 3.

This method permits the quantification of bromates up to a concentration greater than or equal to 5 μg BrO_3^-/l but does not allow the goal of a 1 μg BrO_3^-/l quantification limit to be attained. The direct injection of a larger volume cannot be envisaged for real waters as it would lead to a column overload by the major anions and an enlargement of the peaks. To increase the injected volume, the sample needs to be preconcentrated in a column.

Figure 3 *Injection of a Reference Solution at 5 μg BrO₃⁻/l*

(200 μl loop)

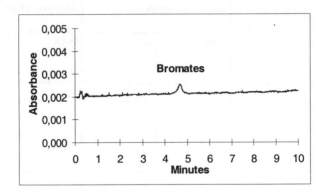

The analysis is therefore carried out in conditions identical to those of the method using conductimetric detection (table 3).[11] The calibration curve obtained in these conditions is presented in figure 4. The chromatograms of a real water containing 5 μg/l obtained by conductimetric and visible detection are presented in figures 5 and 6. On figure 5, the bromate peak shows a lack of symmetry. It makes integration very difficult and quantification imprecise. In fact, if this lack of symmetry is included in the peak, the concentration is 6.0 μg/l and, if it is not, the concentration is 3.8 μg/l. This represents a 37% divergence between the two measured concentrations.

Visible detection after post-column reaction (figure 6) is translated by a chromatogram that only presents 3 peaks. As the chlorides do not react with chlorpromazine, the bromates peak is well isolated. It is perfectly symmetric and the return to the baseline between each peak is total. This simplifies integration.

Consequently, with visible detection the concentration of the analysed water is 5.1 μg/l. This value is the intermediary between the two concentrations obtained using conductimetric detection (3.8 and 6.0 μg/l).

Tests carried out in real waters demonstrated that this method makes it possible to detail the concentration in bromates for matrices presenting a low concentration (2 μg/l), quantify the bromates at a concentration lower than the reference method quantification limit and improve the quantification in waters containing an element resulting in a non-symmetric bromates peak.

Table 3 *Analytical Conditions after Optimisation*

Parameters	Conditions
Analysis column	AS9SC
Guard column	AG9SC
Preconcentration column	AG9SC
Analysis eluant	H3BO3 40 mM / NaOH 20 mM
Purge eluant	H3BO3 250 mM / NaOH 100 mM
Preconcentration	1.6 ml
Eluant flow	2 ml/mn
CLP/HCl mix	1.3 ml/mn
Detection	- Conductimeter - UV / Visible : 525 nm

Figure 4 *Calibration Curve obtained after Post-column Derivatization*

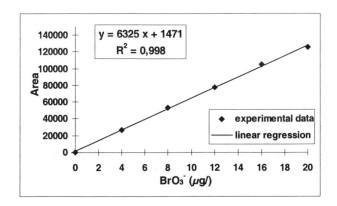

Eight real ozonated and discharged waters were analysed in parallel using the reference method (conductimetry) and the alternative method (table 4). The indicated concentration corresponds to the average of 3 repeated analyses. Table 4 shows the concentration divergences obtained by the alternative method and the reference method are between 0 and 16%. The obtained divergences are probably due to the phenomenon observed in conductimetry (the lower the concentration, the higher the incidence of a non-symmetric bromates peak on the quantification) and allow us to believe that the alternative laboratory method in fact gives the "right" value as the peaks are more easily integrated.

5.2.2   Method Validation. The method validation stages are linearity, limit of detection and limit of quantification, specificity, precision, accuracy and robustness. The use of data relative to the method validation is currently underway.

Figure 5 *Preconcentration of a Real Water containing approximately 5 µg/l of BrO₃⁻ - Conductimetric Detection*

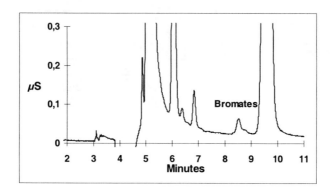

Figure 6 *Preconcentration of a Real Water containing approximately 5 µg/l of BrO₃⁻ - Visible Detection*

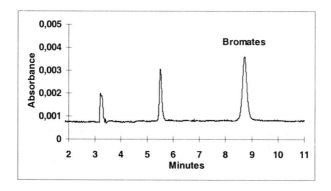

6 CONCLUSIONS

A field method based on the oxidation of chlorpromazine by bromates has been developed, but the overly large number of interfering agents and the difficulties in removing them prevents the use of this method in real waters. The separation of bromates and the interfering agents by ionic chromatography has led to the development of a laboratory method that is more sensitive and less subject to interferences that the current reference method. This new method allows the analysis of ozonated and drinking waters. The method is currently being validated.

Table 4 *Comparison between the Chlorpromazine Method and the Reference Method*

Origin of waters	Bromate Concentration ($\mu g/l$)		Accuracy / IC/CD (%)
	conductimetric Detection	Visible Detection (chlorpromazine)	
Water n°6	2,5	2,1	16
Water n°7	3,7	3,7	0
Water n°8	5,7	5,1	11
Water n°9	6,9	6,1	12
Water n°10	7,0	6,7	4
Water n°11	11,5	10,8	6
Water n°12	14,6	14,2	3
Water n°13	17,5	17,6	1

References

1. S. FONTANA, *Hydroplus*, 1997, **70**, 33.
2. OMS, Directives de qualité pour l'eau de boisson, Recommandations, 1993, Volume 1, Genève.
3. M.H. HASHMI, H. AHMAD, A. RASHID and A.A. AYAZ, *Anal. Chem.*, 1964, **36**, 2028.
4. J.C. MACDONALD and J. YOE, *Anal. Chim. Acta*, 1963, **28**, 383.
5. I. ODLER, *Anal. Chem.*, 1969, **41**, 1116.
6. G. GORDON, B. BUBNIS, D. SWEETIN and C. KUO, *Ozone Sci. Eng.*, 1994, **16**, 79.
7. G. GUOQUAN, J. LI, Q. YANG and W. HUAIGONG, *Analytical letters*, 1993, **26**, 2277.
8. M. DENIS and W.J. MASSCHELEIN, *Analusis*, 1983, **11**, 79.
9. M.A. ABDALLA and H.M. AL-SWAIDAN, *Analyst*, 1989, **114**, 583.
10. H. WEINBERG, *J. Chromatogr. A*, 1994, **671**, 141.
11. M.C. MULLER, B. KOUDJONOU, E. COSTENTIN and P. RACAUD, *in : J.I.E. 94, 11ème Journées Information Eaux*, 1994, 22-1.
12. D.P. HAUTMAN and M. BOLYARD, *J. Chromatogr.*, 1992, **602**, 65.
13. L. CHARLES, D. PEPIN and B. CASETTA, *Anal. Chem.*, 1996, **68**, 2554.
14. J.T. CREED, M.L. MAGNUSON, J.D. PFAFF and C. BROCKHOFF, *J. Chromatogr. A*, 1996, **753**, 261.

ON-LINE MONITORING OF THE RIVER TRENT - TO REDUCE THE LEVEL OF THM FORMATION

J E Upton*, B E Drage* and Dr M Purvis**

*Severn Trent Water Ltd, St Martins Road, Coventry, CV3 6SD,UK
**Severn Trent Laboratories Ltd, Torrington Avenue Coventry, CV4 9GU,UK

1 INTRODUCTION

Severn Trent Water has been investigating the potential of using the River Trent as a drinking water supply resource. Recent semi-drought conditions in the UK over the last few years mean that there is a need to safeguard existing supplies in the East Midlands. The River Trent, hitherto an 'untapped' resource for potable supplies, contains the treated sewage from 4 million people plus Industry in the Midlands and along with diffuse drainage sources means that the river quality can be highly variable.

Severn Trent have established a very advanced on-line monitoring unit on the bankside of the River Trent, to establish the necessary river intake protection to the proposed water supply scheme. The on-line monitors include traditional inorganic and oil monitors in use extensively elsewhere, but in addition the River Trent monitoring unit includes a SAMOS-LC system using a conventional HPLC technique on-line for the analysis of organic pollutants. This monitor originally developed for the analysis of triazine and phenylurea herbicides in the River Rhine system is unique in this River Trent installation in that it includes a modification to allow the detection of several acid herbicides and phenolic compounds known to be frequently present in the River Trent.

An unusually high level of bromide (greater than 300 µg/l) in the River Trent can create potential difficulties in the treatment of this water with the formation of total Trihalomethane (THM) concentrations greater than 100 µg/l in the proposed blend with River Derwent water. THM formation potential bench scale tests have helped determine an upper level of bromide in the raw Trent water at which treatment can proceed in the confidence that the total THM limit will not be exceeded. A further addition to the extensive on-line monitoring suite has been the incorporation of a Bromide Analyser on-line, from which alarms are set to stop abstraction from the river to the bankside storage lakes when the bromide level in the river reaches concentrations greater than 250 µg/l. On-line Dissolved Organic Carbon (DOC) measurement is also available.

This paper demonstrates how these sophisticated on-line monitoring techniques can provide effective river intake protection on what is considered the most significant lowland water treatment challenge in the UK. The first time recorded use of an on-line bromide monitor for drinking water intake protection demonstrates how this technique can safeguard the treatment flow sheet from the risk of high concentrations of disinfectant by-products.

THMs are produced by the reaction of hypohalogen acids and their co-compounds on susceptible organic materials dissolved in water. Many such reactive organics are present in river and reservoir waters. Probably the most prevalent are those based on substituted meta hydroxy aromatic ring structures. Such compounds are readily attacked by halogenating agents at two of the adjacent unsubstituted ring carbon atoms. Further halogen addition and hydrolysis then leads to the breaking of the ring and the formation of intermediate ketone structures. The latter slowly hydrolyse to give a range of THMs plus carboxylic acids. The rate of progress of the latter step is pH sensitive. Transformation of the intermediate compounds is not dependent on the presence of any halogen residual which means that some further THM production may be experienced in the distribution system without further chlorine dosing. Another reactive compound is chlorophyll. In this case the pyrrole group is probably the point of attack.

Hypohalogen acids can be produced by the reaction of the halogen itself with water or the reaction of the halide with an oxidant of higher oxidative potential. This could be a compound such as ozone but, in the case of the effect under study, bromide is oxidised to hypobromous acid by a free chlorine residual.

The amount of THMs produced is affected by:

- temperature
- level of oxidant - in this case free chlorine residual
- amount of dissolved organic carbon (DOC)
- level of bromide present
- contact time
- nature of the organic compounds present
- pH

Examination of the published literature[1] reveals many differing approaches and formulae for the calculation of THM production and speciation involving multivariable matrices incorporating some or all of these factors. However the time and resources available would not permit a full exploration of the variables. It was decided therefore to follow as closely as possible the procedures likely to be carried out during treatment at Church Wilne WTW and to try to establish a relationship between the DOC and bromide levels and THM formation for the range of values likely to be encountered with an understanding that such a relationship may not apply outside these values. Where practicable 'worst scenario' conditions were chosen.

Following Severn Trent Water's decision to use the River Trent as a future water resource the catchment has been closely monitored and bromide was found in considerably higher concentrations than for 'normal' lowland rivers in the UK.

Bromide is not considered a toxic constituent of river systems and therefore it is not a regulated parameter in the River Trent. However, the origin of the bromide sources are important to this abstraction strategy.Monitoring of the River Trent catchment has shown these bromide inputs to originate from the highly industrialised discharges in Birmingham and consequently the River Tame.

Directly parallel to this work, a request was made to Severn Trent Laboratories (STL), already charged with operating the on-site monitoring facilities, to provide timely bromide and DOC measurements to a specification set by Severn Trent Water.

2 ON-LINE MONITORING REQUIREMENTS

With monitoring for bromide, one problem is the need for specificity, especially as comparatively high chloride levels will be present.

Cost efficiency (both for the capital equipment purchase and the running costs) together with high reliability, low operator involvement and the ability to generate and transmit alarms were of equal importance.

Ion chromatography (IC) was the obvious choice of technique as the anion of interest is separated chromatographically from likely interferences. Dionex are amongst the world leaders in this technique and indeed produce a specific on-line monitor as a robust field unit. This is based on IC and a single chemistry, single channel, single calibration unit is available at around £70K including sampling and signal outputs.

The option of adapting a laboratory based instrument was also considered as STL were already operating on-line HPLC based on a laboratory unit very successfully. The "friendly" site environment which includes heating, air-conditioning and experience gained in the establishment of telemetry links and remote control via PC Workstations with modem facilities with standard laboratory instruments supported the choice of this second option. An on-line filtered sample is also available thereby simplifying problems of sample delivery. Potential cost savings were hence identified in customising a laboratory ion chromatograph to fulfil the needs of a bromide monitor operating on-line.

Sample delivery from the constant head containing ultrafiltered water was achieved using a PEEK sample line (i.d. 0.76 mm) coupled to a Watson Marlow peristaltic pump (model no. 313S) pumping at 0.8 ml/min. This pumped river water constantly through a sample loop to waste. Injection of sample was achieved by automatic switching in-line of the loop. A microbore pumping system was utilised to conserve eluent and calibration was achieved manually by switching a toggle to select either sample stream or calibrant solution. Calibration was single point and matched to the river water ranges. Fluoride, chloride, bromide, nitrate, phosphate and sulphate were calibrated at 10, 200, 1, 100, 10 and 200 mg/l respectively.

Anion separation was achieved using an AS-14 column with carbonate/bicarbonate eluent at 25 $^\circ$C and a flow rate of 0.3 ml/min. The analysis took 16 minutes and used electrical conductivity detection. The monitor required visits three times a week of around 45 mins each for the necessary calibrations and system maintenance and is viewed from the central laboratory facility (some 80 km distant) on a daily basis via remote access software. Customised software is used to effect contact closures to telemetry and a dial out feature to a radiopager can also be used. The cost was substantially lower to the custom built unit, at a total cost of £32K. The system is proving to be very reliable with >95% operating availability and is set to sample river water on a 2-hourly cycle.

DOC is measured using a commercially available on-line filtered TOC (total organic carbon) monitor. This is based on a filtered sample being acidified to remove inorganic carbon. Oxidation of the organic carbon is achieved by addition of persulphate and the action of UV light. The carbon dioxide produced is measured by infra-red spectroscopy. Connection into telemetry systems is a standard feature and allows alarming requirements to be met. These two monitoring systems allow the necessary information to be available to protect against the intake of Trent water highly susceptible to THM formation.

3 EXPERIMENTAL WORK

The bench scale testing to assess the impact of treating Trent water involved two separate exercises, a DOC/bromide THM matrix study and a series of Total THM (TTHM) formation potential trials, being carried out to determine the relationship between DOC, bromide levels and the formation of THMs. The experiments were designed in the knowledge of bromide concentrations in the Trent varying between 100 and 800 µg/l at the point of abstraction.

3.1 DOC/Bromide THM Matrix Study

3.1.1 Operational conditions used. A temperature of 22 °C was selected as being representative of summer temperature for the water passing through the distribution system. All samples were incubated at this temperature for the duration of the tests.
The chlorine levels were problematical as with batch systems it is not possible to establish a steady state. However using frequent monitoring and adjustment, an approximate residual of 1 mg/l of free chlorine was maintained for the first 2 hours. The residuals were adjusted after 24 hours to give an approximate residual of 0.5 mg/l free chlorine after 48 hours.

The raw river waters were coagulated and filtered through GFC circles (1.2 µm pore size) before use and a borehole water was selected for blending being very low in DOC and bromide. River and borehole waters were blended to give a range of DOC values and each of these was spiked to give a range of bromide concentrations up to 800 µg/l. Samples were submitted for each DOC dilution. All combinations of materials were also submitted for bromide analysis.

Samples for THM analysis were submitted after 48 hours contact time. It was thought that this would give the most likely time period during which Church Wilne treated water would be consumed.

Organic compounds found in water give a wide variety of properties regarding THM formation potential. For example humic acids are quoted as being much more reactive in both rate and total potential than fulvic acids. It seemed likely therefore that water from the River Derwent with a considerable contribution from the moorland in the Derwent Valley would have a different THM formation profile to the River Trent with its more industrial and sewage derived organics. Trials were therefore conducted on both river waters.
No pH adjustment was carried out. The pH of the coagulated water was approximately 7.1 and the borehole about 7.4. All the waters had relatively high hardness content and were therefore fairly well buffered.

3.1.2 Results. The results are plotted in Figures 1 and 2 for the Derwent and Trent river waters respectively. These figures show visually the matrix effect of DOC, bromide and TTHM production. This information can also be used to enable calculation of critical bromide levels at different DOC concentrations by plotting DOC*Br against TTHM production and developing equations for the relationship between them. This relationship is quoted in literature[1] where bromide is the TTHM limiting factor.

The plot shows a relationship for instance on the River Trent exercise of the 48 hour TTHM and DOC*Br of $y = 0.1336x^{0.5016}$ with an r^2 of 0.9399 and for the River Derwent

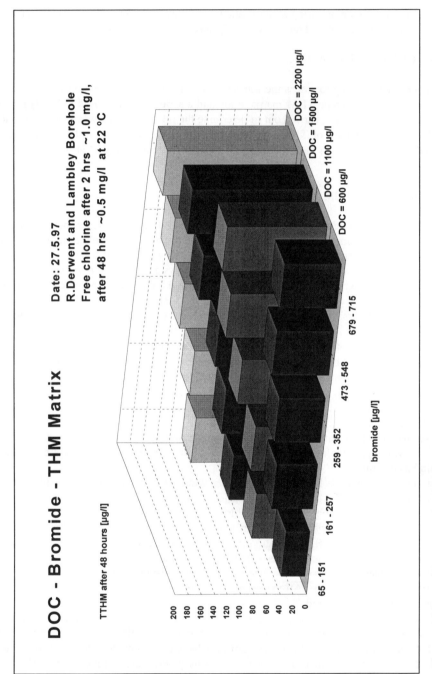

Figure 1 DOC-Bromide-THM Matrix with River Derwent water

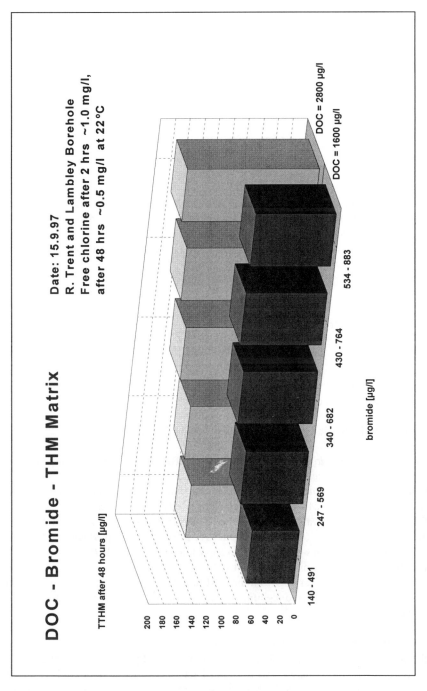

Figure 2 DOC-Bromide-THM Matrix with River Trent water

$y = 0.2009x^{0.4889}$ with an r^2 of 0.9592 where y is the TTHM (μg/l) and x is the product of the bromide in μg/l and the DOC in μg/l.

A number of scenarios can be produced from this by selecting a range of DOC values in the treated water from each source and a range of blends. For instance other work has shown that the DOC on a 1:1 blend of the two sources (the maximum Trent water likely to be used) could generally achieve a DOC level at the entry to the contact tank of approximately 2000 μg/l.

TTHM $= 0.2(2000[Br])^{0.5}$ μg/l for River Derwent water and

TTHM $= 0.13(2000[Br])^{0.5}$ μg/l for River Trent water

The mean summer level of bromide in the River Derwent is about 150 μg/l which currently causes negligible TTHM limit exceedances with the current treatment regime.

To achieve the maximum TTHM of 100 μg/l under these circumstances the maximum acceptable concentration of bromide in the River Trent water would be given by

$$100 = \frac{50*0.13(2000*Br)^{0.5}}{100} + \frac{50*0.2(2000*150)^{0.5}}{100}$$

$$200 = 0.13(2000*Br)^{0.5} + 110$$

$$Br = 240 \ \mu g/l$$

Having taken the 'worst case' treatment conditions a safe alarm level for Trent bromide at normal final water DOC concentrations can be predicted to be 250 μg/l.

3.2 Total THM formation potential of 50:50 Trent : Derwent water

3.2.1 Operational Conditions Used. (see Table 1) Total THM formation potentials were assessed at both ambient temperature and 22 $^{\circ}$C

As above initial chlorination was for two hours, with the aim of maintaining 1 mg/l free chlorine. Targets for the 24 and 48 hour residuals were 0.3 to 0.5 mg/l.

The water under test was a 50/50 blend of River Trent and River Derwent water. This was treated by coagulation (with and without PAC), filtration and passage through an aged GAC column. One stream was enhanced to approximately 500 μg Br/l.

As before 48 hour reaction times were used.

The pH of the water emerging from the GAC columns was generally 7.6 to 7.9. On the rare occasions where it differed significantly from this, lime was used to adjust to within this range.

3.2.2 Results. The results are plotted in Figure 3.

For the control conditions A the TTHM results were generally at or below 100 μg/l prior to June with bromide levels below 350 μg/l and DOC levels below 3100 μg/l. For July to September exceedances occurred when temperatures were >16 $^{\circ}$C, DOC >2000 and Br >250 μg/l. These results confirm the earlier work giving a 250 μg Br/l limit for a DOC

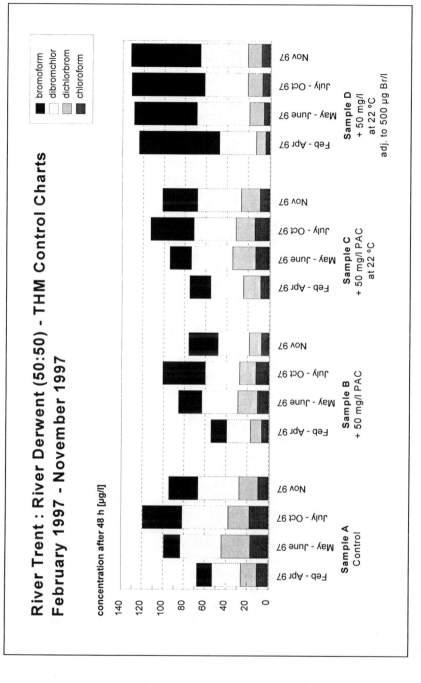

Figure 3 River Trent : River Derwent (50:50) - THM Control Charts

Table 1 *Four different conditions from the variables examined.*

Sample nation	Desig-	PAC 50 mg/l Norit SA	Br adjusted to 500 μg/l	Plant temp	Temp adjusted to 22°C
A	-	-	✓	-
B	✓	-	✓	-
C	✓	-	-	✓
D	✓	✓	-	✓

level of 2000 μg/l.

For condition B the TTHM results were at or below 100 μg/l with a few exceptions where bromide level exceeded 300 μg/l. This shows that 50 mg/l PAC extends the Br level of tolerance by reducing the DOC (from~3000 to ~2000 μg/l.) before chlorination.

Condition C data shows that the higher temperature (22 °C) increased the total THM results for February to June by ~20% and by ~7% for the period July to September. In general terms the increase in TTHM production caused by adjusting the temperature to 22 °C cancels out the effect of 50 mg/l of PAC.

Considering condition D very few TTHM results are below 100 μg/l. Bromide levels greater than 500 μg/l cannot be tolerated for any DOC level used in these tests (2000 to 2500 μg/l.) at 22 °C. Comparing the individual THMs in Figure 3 the increase in bromoform concentration for condition D emphasises the effect that increased bromide has on the shift towards brominated THMs. The oxidation by hypochlorite of organics/bromide and halogenation (substitution) by the resulting oxidation product hypobromite gives rise to the shift. The findings reported by Rook and co-workers[2] are clearly seen here, notably that the reaction of hypochlorite is mainly oxidative whereas that of hypobromite is predominantly that of an halogenating agent. The resultant preferential incorporation of bromide does of course increase the measured yield of TTHM due to the relatively higher atomic mass of bromine when compared to that of chlorine.

4 DISCUSSION

The introduction of high bromide/DOC Trent water into the River Derwent based raw water system at Church Wilne can be screened out by the on-line monitors as described earlier.

Considering the potential of high numbers of algae influencing THM levels, a number of preventative measures have been taken. The bankside storage (3 lakes in series) is to be managed to minimise algae at the draw off point by allowing growth in Lake 1, strategic use of barley straw and development of integrated wetlands. The exit of lake 3 will be monitored continuously for chlorophyll "a" with appropriate alarms set.

The measures described allow use of Trent water during favourable conditions. In order to make continuous use of this resource at Church Wilne WTW the following issues require investigation and action;

- continue to monitor high bromide discharges currently entering the Trent catchment system.

- review chlorination regimes used and disinfection treatment options available for Church Wilne WTW.
- consider using Upper Trent water as an alternative resource. This can potentially be transferred to the bankside storage lakes via the Trent and Mersey canal.
- investigate alternative full treatment strategies for longer term use of River Trent water.

5 CONCLUSIONS

High bromide Trent water is likely to lead to THM exceedances even when blended with Derwent water during periods of high temperatures and DOC levels.

On-line monitoring of bromide and DOC has been installed to protect against intake of Trent water highly susceptible to THM formation.

Combinations of PAC and GAC can be used to reduce DOC, THM precursors and THMs formed by in-process chlorination. This allows flexibility of monitor shut off settings to be balanced against cost effective treatment.

Future use of alternative treatment strategies and/or point of intake need to be investigated to allow long term continuous use of the River Trent as a drinking water resource.

References

1.R E Rathbun, 'Bromine incorporation factors for trihalomethane formation for the Mississippi, Missouri, and Ohio Rivers, The Science of the Total Environment 192 (1996) 111-118.
2.J J Rook et al, 'Bromide Oxidation and Organic Substitution in water treatment', *J. Environ. Sci. Health,* A13(2), 91-116 (1978)

Standards and Regulation

DBP STANDARDS AND REGULATIONS: THE NORTH AMERICAN SITUATION

S. Regli, M. Cox, T. Grubbs, and J. McLain

Office of Ground Water & Drinking Water
US Environmental Protection Agency
401 M St. SW
Washington DC 20460

1 INTRODUCTION

The purpose of this paper is to present an overview of disinfection byproduct (DBP) standards for drinking waters as they exist in North America, and as such standards are further being developed in the United States (US). The primary focus is the status of DBP regulation development within the US, associated issues, and the process through which key issues are being resolved.

2 OVERVIEW OF EXISTING STANDARDS AND REGULATIONS

2.1 Regulations in the United States

US Environmental Protection Agency (EPA) set an interim maximum contaminant level (MCL) for total trihalomethanes (TTHMs) of 0.10 mg/l as an annual average in November 1979 (USEPA, 1979).[1] The purpose of this regulation was to lower exposure from THMs and other byproducts formed from chlorination.

The interim TTHM MCL applies to any public water system (PWS) serving at least 10,000 people that add a disinfectant to the drinking water during any part of the treatment process. At their discretion, States may extend coverage to smaller PWSs. However, most States have not exercised this option.

EPA chose to limit applicability of the TTHM standard to the largest systems for a variety of reasons. The vast majority of small PWSs are served by ground water that is generally low in THM precursor content and which would be expected to have low TTHM levels even if they disinfect. In addition, smaller systems were considered more vulnerable to increases in microbial risk if they were to make treatment changes to comply with a new TTHM standard. Smaller systems were considered to have less financial resources and technical expertise available for THM control and the majority of waterborne disease outbreaks attributed to inadequate disinfection had occurred in smaller systems. Finally, greater than 80% of the US population receiving disinfected water were served by systems serving greater than 10,000 people.

Compliance with TTHM MCL is based on a running annual average, determined quarterly, of the TTHM concentrations measured at four representative points in the distribution system, i.e., based on 16 samples per year. State agencies, which review system compliance with drinking water standards, may allow for lower monitoring frequencies depending upon the levels of TTHMs detected and the type of source water. Surface water systems may be allowed as few as one sample per quarter at distribution system locations representing maximum residence time. Ground water systems may be allowed as few as one maximum TTHM formation potential measurement per year to determine compliance.

2.2 Canadian and Central AmericanGuidelines

In 1993 Health Canada changed the TTHM guideline from a maximum of 0.35 mg/l to a running annual average of 0.10 mg/l based on quarterly monitoring. The new guideline is designated as interim until such time as the risks from other disinfection byproducts are ascertained. Health Canada does not expect that all water systems will be able to meet the revised TTHM guideline immediately and advise that when water systems are expanded or upgraded, every effort should be made, not only to meet the revised guideline but to reduce concentrations of THMs to as low as possible.

Central American countries use the WHO guidelines in considering control for disinfectants and DBPs. The WHO (World Health Organization) guidelines include values for chlorine and chloramines (when used as disinfectants),bromoform, dibromochloromethane, bromodichloromethane, chloroform, and chloral hydrate. Provisional guideline values are provided for chlorite, bromate, dichloroacetic acid, and trichloroacetic acid. [2] WHO recommends that while considering guideline values for DBPs primary consideration be given to ensuring that disinfection to control for pathogens is never compromised.

3 PROPOSED DBP REGULATIONS IN US

3.1 Background

When the 1979 TTHM standard was promulgated, EPA indicated its intent to revise this standard, possibly lowering the MCL and extending the coverage to all system sizes, as more information and experience with implementing the regulation became available.[1]

3.1.1 Statutory Requirements. The Safe Drinking Water Act (SDWA or the Act), as amended in 1986, further influenced EPA to proceed with revising regulations for disinfection byproducts (DBPs). The 1986 SDWA Amendments required EPA to publish a maximum contaminant level goal (MCLG) for each contaminant which, in the judgement of the USEPA Administrator, may have any adverse effect on the health of persons and which are known or anticipated to occur in public water systems (Section 1412(b)(3)(A)). MCLGs are to be set at a level at which no known or anticipated adverse effect on the health of persons occur and which allows an adequate margin of safety (Section 1412(b)(4)).

The Act also requires that at the same time USEPA publishes an MCLG, which is a non-enforceable health goal, it also must publish a National Primary Drinking Water Regulation (NPDWR) that specifies either a maximum contaminant level (MCL) or treatment technique (Sections 1401(1) and 1412(a)(3)). MCLs are to be set at levels for which the Best Available Technology (BAT) can achieve, taking costs into consideration. USEPA is authorized to promulgate a NPDWR that requires the use of a treatment technique in lieu of establishing a MCL, if the Agency finds that it is not economically or technologically feasible to ascertain the level of the contaminant . For example, since it is not feasible to measure *Giardia* in the finished water, EPA has a treatment technique that measures the effectiveness through turbidity and disinfection practice.

Recognizing that many new DBPs had been identified as frequently occurring in drinking waters (e.g., haloacetic acids, chloral hydrate, haloacetonitriles), EPA intended to regulate new DBPs while also revising the existing TTHM standard. [3]

3.1.2 Regulatory Issues Concerns with how to address several major issues delayed EPA s decision to propose new DBP standards. [3] One issue concerned potential trade-offs between microbial and DBP risks. This issue raised questions regarding uncertainties in defining microbial and DBP risks, levels of risks that would be considered acceptable and at what cost, and defining practical (implementable) criteria to demonstrate that an achievable risk had been reached. A related issue was integration with the Surface Water Treatment Rule (SWTR) which had been promulgated in 1989. Although the SWTR only mandated 3-log (99.9%) removal or inactivation of *Giardia* and 4-log (99.99%) of viruses, EPA guidance recommended higher levels for poorer quality source waters. EPA was concerned that systems would reduce microbial protection to levels nearer to the regulatory requirements by reducing disinfection and possibly increase microbial risks in an effort to meet DBP MCLs.

A third issue was the use of alternate disinfectants to limit chlorination byproducts. The Agency recognized that while alternate disinfection schemes (e.g., ozone and chloramines) could greatly reduce byproducts typical of chlorination, little was known about the byproducts of the alternate disinfectants and their associated health risks. EPA did not want to promulgate a standard that encouraged major shifts to alternate disinfectants unless the associated risks (including both those from byproducts and differential microbial risks from a change in disinfectants) were adequately understood.

A fourth issue was best available technology. How BAT was defined would determine the levels at which MCLs were set. The levels at which MCLs were set would influence the types of technologies that systems would choose to comply with (systems are not required to use BAT to comply and would tend to use the least cost technology assuming there were no other regulatory constraints). For example, defining BAT with use of alternate disinfectants to chlorine would drive the chlorination byproduct MCLs down; this could result in significant increased exposure to (not well characterized) alternate byproducts. Similarly , defining BAT as including an advanced DBP precursor removal technology, such as deep bed granular activated carbon or membrane technology, with chlorine, could also lead to a low MCL for chlorinated DBPs; and because of lower costs for meeting such an MCL through use of alternative disinfectants (e.g., ozone and chloramines), large industry shifts in disinfection practice might result without knowing the risk trade-offs.

3.1.3 Negotiated Rule Making In 1992 USEPA initiated a negotiated rulemaking to assist in developing a disinfectants/disinfection byproducts (D/DBP) rule.[3] The negotiators (referred to as the Committee) included representatives of State and local health and regulatory agencies, public water systems, elected officials, consumer groups and environmental groups.

Early in the process, the Committee agreed that large amounts of information necessary to understand how to optimize the use of disinfectants to concurrently minimize microbial and DBP risk on a plant-specific basis were unavailable. One of the major goals of the Committee was to develop an approach that would reduce the level of exposure from disinfectants and DBPs without undermining the control of microbial pathogens. The intention was to ensure that drinking water is microbiologically safe at the limits set for disinfectants and DBPs and that these chemicals do not pose an unacceptable risk at these limits.

Following months of intensive discussions and technical analysis, the Committee recommended the development of three sets of rules: a two-staged D/DBP Rule, an interim Enhanced Surface Water Treatment Rule (IESWTR), and an Information Collection rule (ICR).[3]

   - **D/DBP Rules.** The first Stage D/DBP rule would substantially reduce exposure from DBPs at low increased household costs for most systems. This wold not influence major shifts toward use of alternative disinfectants to chlorine. A second Stage D/DBP rule that achieved additional reductions in DBP exposure would later be proposed when more occurrence, health effects, and treatment information became available.

   - **Enhanced Surface Water Treatment Rules.** The IESWTR, which would only apply to systems serving at least 10,000 people, would increase control for *Cryptosporidium* and prevent increases in microbial risk while systems complied with the Stage 1 D/DBP rule. The IESWTR included several rule options, one or more of which would be later selected for final promulgation when pertinent information became available. A long-term ESWTR (LTESWTR) would later be proposed and promulgated that would apply to systems serving 10,000 people or less and prevent increases in microbial risk while small systems complied with the Stage 1 D/DBP rule; the LTESWTR could also include additional refinements for larger systems.

   - **ICR.** The ICR would provide data from which to select the best IESWTR option(s) and support the development of the Stage 2 D/DBP rule.

A ground water disinfection rule, to be developed later, would prevent increases in microbial risk while ground water systems complied with the Stage 1 D/DBP rule. Compliance dates for the Stage 1 D/DBP rule would coincide with the compliance dates for the microbial standards.

3.2 1994 Proposed D/DBP Rule

The 1994 USEPA proposed a two staged regulation for DBPs, an IESWTR, and the ICR as recommended by the Committee. Key elements of the Stage 1 D/DBP rule are described below.[3]

   3.2.1   MCLGs/MCLs/MRDLGs/MRDLs   EPA proposed MCLGs of zero for chloroform, bromodichloromethane, bromoform, bromate, and dichloroacetic acid and MCLGs of 0.06 mg/L for dibromochloromethane, 0.3 mg/L for trichloroacetic acid, 0.04 mg/L for chloral hydrate, and 0.08 mg/L for chlorite. In addition, EPA proposed to lower the MCL for TTHMs from 0.10 to 0.080 mg/L and added an MCL for five haloacetic acids (i.e., HAA5 - the sum of the concentrations of mono-, di-, and trichloroacetic acids and mono- and dibromoacetic acids) of 0.060 mg/L. Compliance monitoring with the TTHM and HAA5 MCLs would be similar to the existing TTHM standard but include new provisions for reduced monitoring depending upon source water and system size.

EPA also, for the first time, proposed MCLs for two inorganic DBPs: 0.010 mg/L for bromate and 1.0 mg/L for chlorite. Compliance with the bromate MCL, which would only pertain to systems using ozone, would be based on a running annual average determined monthly based on monthly measurements at the entry to the distribution system. Compliance with the chlorite MCL, which would only pertain to systems using chlorine dioxide, would be determined monthly (because of health risks associated with short term exposure) based on three samples taken in the beginning, middle, and end of the distribution system.

EPA also proposed maximum residual disinfectant level goals (MRDLGs) of 4 mg/L for chlorine and chloramines and 0.3 mg/L for chlorine dioxide; and maximum residual disinfectant levels (MRDLs) for chlorine and chloramines of 4.0 mg/L, and 0.8 mg/L for chlorine dioxide. MRDL compliance monitoring for chlorine (measured as free or total chlorine) and chloramines (measured as total chlorine) would be based on an annual average of monthly sampling within the distribution; the frequency and location of sampling would coincide with that required for coliform monitoring and depend on population served. MRDL compliance monitoring for chlorine dioxide would be based on daily samples prior to entry into the distribution system with no daily sample allowed to exceed the MRDL. The more stringent monitoring requirements for chlorine dioxide are due to concern of neuro-developmental health risks.

3.2.2 Best Available Technologies   EPA identified the best available technology (BAT) for achieving compliance with the MCLs for both TTHMs and HAA5 as enhanced coagulation or treatment with granular activated carbon with a ten minute empty bed contact time and 180 day reactivation frequency (GAC10), with chlorine as the primary and residual disinfectant. It was estimated that most systems using surface water would be able to achieve the TTHM and HAA5 MCLs through use of the BAT. Since enhanced coagulation would also be required as a treatment technique (see below) most systems would not elect to use alternative disinfectants for achieving the MCLs.

The BAT for achieving compliance with the MCL for bromate was control of ozone treatment process to reduce formation of bromate. The BAT for achieving compliance with the chlorite MCL was control of precursor removal treatment processes to reduce disinfectant demand, and control of chlorine dioxide treatment processes to reduce disinfectant levels. EPA identified BAT for achieving compliance with the MRDL for chlorine, chloramine, and chlorine dioxide as control of precursor removal treatment processes to reduce disinfectant demand, and control of disinfection treatment processes to reduce disinfectant levels.

3.2.3 Treatment Technique   EPA proposed a treatment technique that would require surface water systems and groundwater systems under the direct influence of surface water that use conventional treatment or precipitative softening to remove DBP precursors by enhanced coagulation or enhanced softening. A system would be required to remove a certain percentage of total organic carbon (TOC) (based on raw water quality) prior to the point of continuous disinfection as indicated in Table 1.

Table 1: *Required Removal of TOC by Enhanced Coagulation/Enhanced Softening for Surface-Water Systems Using Conventional Treatment*

Source water TOC mg/l	Source water alkalinity mg/l as $CaCO_3$		
	0-60a	>60-120a	>120a
>2.0-4.0	40.0%	30.0%	20.0%
>4.0-8.0	45.0%	35.0%	25.0%
>8.0	50.0%	40.0%	30.0%

a Exceptions allowed as approved by State if system demonstrates through jar testing

EPA also proposed a procedure to be used by a PWS not able to meet the percent reduction, to allow them to comply with an alternative minimum TOC removal level based on a point of diminishing returns with respect to alum added and TOC removed (defined when less than 0.3 mg/l TOC is removed per 10 mg/l alum dose applied). Compliance for systems required to operate with enhanced coagulation or enhanced softening would be based on a running annual average.

3.2.4 Preoxidation (Predisinfection) Credit The proposed rule did not allow PWSs to take credit for compliance with disinfection requirements in the SWTR/IESWTR prior to removing required levels of precursors unless they met specified criteria. This provision was to encourage systems to delay application of the disinfectant until after substantial removal of DBP precursors. It was anticipated that under the IESWTR, systems would provide, where appropriate, additional inactivation or removal to compensate for loss of disinfection prior to precursor removal. Ref

3.2.5 Analytical methods EPA proposed nine analytical methods (some of which can be used for multiple analyses) to ensure compliance with proposed MRDLs for chlorine, chloramines, and chlorine dioxide. EPA proposed methods for the analysis of TTHMs, HAA5, chlorite, bromate and total organic carbon.

4 1994 PROPOSED IESWTR

The proposed IESWTR included several rule options intended to increase protection from Cryptosporidium and simultaneously prevent increased microbial risk while systems complied with the Stage 1 D/DBP rule. [4] EPA also indicated intent to later issue a Notice of Data Availability (NODA) when more data became available, describing which of the proposed rule options it considered most appropriate to promulgate and to solicit public comment.

5 RECENT DEVELOPMENTS CONCERNING DBP REGULATION

5.1 1996 SDWA Amendments

In 1996, the US Congress reauthorized the Safe Drinking Water Act and as part of these provisions required EPA to promulgate a Stage 1 disinfectants/ disinfection byproducts (D/DBP) rule by November 1998, an LTESWTR by November 2000, a GWDR by May 2002, and a Stage 2 D/DBP rule by May 2002.

5.2 1997 Federal Advisory Committee and NODAs

To help meet the statutory deadlines for the IESWTR and Stage 1 D/DBP rule and to maximize stakeholder participation, the Agency established an Advisory Committee to collect, share, and analyze new information and data, as well as to build consensus on the regulatory implications of this new information. The Committee also considered the effect of not having pathogen occurrence data from the ICR to help select proposed IESWTR options. The Committee reached agreement on the following major issues: 1) maintaining the proposed MCLs for TTHMs, HAA5 and bromate; 2) modifying the enhanced coagulation requirements as part of DBP control; 3) allowing credit for compliance with applicable disinfection requirements should continue to be allowed for disinfection applied at any point prior to the first customer, consistent with the existing SWTR; 4) including a microbial benchmarking/profiling procedure to provide a methodology and process by which a PWS and the State, working together, will assure that is no significant reduction in microbial protection as the result of modifying disinfection practices in order to meet DBP MCLs; 5) modifying the turbidity performance requirements and monitoring requirements for individual filters; 6) issues related to the MCLG for *Cryptosporidium*; 7) requirements for removal of *Cryptosporidium*; and 8) provision for conducting sanitary surveys.

Information on the above first three elements were discussed in the NODA pertaining to the Stage 1 D/DBP rule [5] and information on the latter five elements were discussed in the NODA pertaining to the IESWTR [6]. Elements two and three are discussed briefly below.

5.2.1 Key Enhanced Coagulation/Enhanced Softening Requirements

At the time of proposal it was not yet apparent how feasible it would be for most systems to comply with the TOC removal requirements. A substantial amount of data has been collected since the time of proposal to evaluate the proposed criteria [5].

The technical working group (TWG) supporting the Committee believed that 90 percent of systems with source waters with TOC >4.0 mg/L (6 of the 9 boxes in the 3 x 3 matrix -see Table 1) would be able to meet the 1994 proposed TOC removal requirements. However, the Committee recommended that waters with raw-water specific UV absorbance (SUVA) ≤ 2.0 L/mg-m be given an exemption to enhanced coagulation and enhanced softening. SUVA is an indicator of the humic content of a water. Since coagulation removes primarily humic matter, waters with low-SUVA values contain primarily nonhumic matter not amenable to enhanced coagulation. Accordingly, in the final Stage 1 D/DBP rule, EPA intends to allow systems to use a raw water SUVA < 2.0 liter/mg-m as a criterion for not practicing enhanced coagulation or enhanced softening.

For waters with TOC >2.0-4.0 mg/L, the Committee recommended that the TOC removal requirements be lowered to 35, 25, and 15 percent for low-, moderate-, and high-alkalinity waters, respectively. The TWG believed that 90 percent of the systems treating low-TOC waters with raw-water SUVA >2 L/mg-m will be able to comply with the revised TOC removal levels.

The TWG also recommended another criterion for demonstrating alternative TOC removals. At the point when the settled-water SUVA is ≤ 2.0 L/mg-m, the residual TOC is mainly composed of nonhumic matter that is not amenable to enhanced coagulation; therefore, it is not productive to add additional coagulant. Because oxidants can destroy

UV, but not TOC, SUVA must be determined on water that has not been exposed to oxidants. Thus, using a settled-water SUVA $\leq$ 2.0 L/mg-m as a point of diminishing returns should be done on jar-tested water (as the slope criterion is done) unless the full-scale plant is not using preoxidation/predisinfection. The TWG believed that these revised requirements will result in a limited amount of transactional costs for the PWSs and States. Accordingly, EPA intends to include this provision in the final rule.

5.2.2 *Disinfection Credit*   TWG analyses indicated that most PWSs, using enhanced coagulation or enhanced softening as required, would be able to meet MCLs of 0.080 mg/L and 0.060 mg/L for TTHM and HAA5, respectively, while maintaining existing disinfection practice. This analysis also indicated that significant precursor removal and DBP reduction can still be achieved with predisinfection left in place. Although in most cases the reduction in DBP formation is not as great as would be accomplished in moving the point of disinfection to after enhanced coagulation, the Advisory Committee recommended balancing the need to maximize precursor removal against the need to substantially maintain existing levels of microbial protection that is provided by many plants through predisinfection. However, for some PWSs that predisinfect just prior to coagulant addition (e.g., rapid mix), significant additional DBP reduction could be achieved without sacrificing disinfection credit by moving the point of disinfectant addition to just after the point of coagulant addition.

The Committee recommended that PWSs continue to receive credit for compliance with applicable disinfection requirements for disinfectants applied at any point prior to the first customer consistent with the existing provisions of the SWTR.

EPA will develop guidance on the uses and costs of oxidants that control water quality problems (e.g., Asiatic clams, zebra mussels, iron, manganese, algae, taste and odor) and whose use will reduce or eliminate the formation of DBPs of public health concern.

5.3 1998 NODA on DBP Health Effects

At the time of this writing EPA is about to issue another NODA pertaining to recent DBP health effects information. Key issues related to the Stage 1 D/DBP rule that are addressed in this Notice include the establishment of Maximum Contaminant Level Goals for chloroform, dichloroacetic acid, chlorite, and bromate and the Maximum Residual Disinfectant Level Goal for chlorine dioxide; and analysis of new epidemiology studies pertaining to risks from cancer and reproductive effects. In the NODA EPA will solicit public comment on the relevance of this information affecting the proposed rule. EPA intends to promulgate the Stage 1 DBP rule in November of 1998 concurrently with the IESWTR to prevent increases in microbial risk. [5, 6]

5.4 Stage 2 D/DBP Rule Development

There are several remaining issues that EPA is attempting to address with additional research and data collection for the Stage 2 D/DBP rule in collaboration with other groups such as the National Toxicology Program, American Water Works Association Research Foundation, and the Centers for Disease Control and Prevention. [7] These issues include:

-       Magnitude of risk from chlorinated waters
-       Relative risk from use of chlorine versus alternative disinfectants

-       Relative risk from brominated versus chlorinated species
-       Characterization of DBP occurrence especially from alternative disinfectants
-       Effectiveness of GAC/membranes to remove DBP precursors
-       Balancing pathogen and DBP risks

Determining the magnitude of the health risks from chlorinated DBPs is important because of the large number of systems in the U.S. that use chlorine as their primary or secondary disinfectant. Understanding the magnitude of the health risks from chlorinated water will be useful when evaluating the different regulatory options for the Stage 2 DBP rule and for evaluating the risk-risk tradeoffs between controlling for microbial pathogens and DBPs.

The second issue is important because using alternative disinfectants may be much more cost effective than using advanced technologies such as membranes or granular activated carbon to comply with MCLs for chlorinated DBPs. On the other hand, alternative disinfectants to chlorine create their own DBPs of concern that need to be considered (e.g., bromate with the use of ozone and chlorite with the use of chlorine dioxide). Since ozone and chlorine dioxide are more effective for inactivating *Cryptosporidium* than chlorine it becomes especially important to understand the health risks of DBPs formed using different disinfectants. However, before these issues can be resolved it is critical that the risks from the byproducts formed from alternative disinfectants be evaluated.

The third issue may be critical because recent evidence suggests that brominated DBPs may be more potent than chlorinated DBPs. This may be important because research indicates that precursor removal technologies (e.g., enhanced coagulation) cause a relative shift from more chlorinated DBPs (e.g., chloroform) to more brominated DBPs (e.g., bromoform). If these two facts are true then the risk reduction achieved with enhanced coagulation may not be as significant as originally assumed. Another important need is to better characterize the occurrence/exposure to the brominated versus chlorinated species to assist in characterizing the risk.

The fourth issue is important because over 50% of organic halides produced by chlorine disinfection are currently unknown. A large portion of these are thought to be non-volatile which are difficult to analyze by traditional DBP methods designed for volatile DBPs such as THMs; polar which are difficult to extract from water; and/or high molecular weight. Oxidation (disinfection) also produces many non-halogenated DBPs. In addition, more information is necessary on the DBPs formed using alternative disinfectants. While there are several research projects in this area, additional work is needed to better quantify the levels of the DBPs from alternative disinfectants.

The fifth issue is critical because to reach lower levels than the proposed MCLs of 80 ug/l for TTHMs and 60 ug/l for HAA5 it may be necessary to use advanced precursor removal technologies such as GAC, ozone and biologically active filtration, or membranes. Raw water quality, existing DBP levels, plant configuration and flow rate, capital costs, and operation and maintenance costs will influence which technology an utility selects. The objective of the DBP treatment research is to optimize the precursor removal capabilities of these technologies and to determine their cost effectiveness for different source water types and system sizes.

Finally, as discussed above, it is critical that a better understanding of the risk trade-offs between microbial and DBP risk is developed. The ability to determine the proper balance between controlling risks from pathogens and DBPs requires large amounts of data, some of which will be developed as part of the current research. Research is under way to better characterize occurrence, treatment efficiencies, and risks associated with exposure from pathogens in drinking water. [7] Since the occurrence of pathogens and DBPs in drinking waters vary greatly, it is very difficult to estimate total national risks or total national risk reductions that might result from different regulatory options. Also, regulatory decisions which lead to minimum social costs at the national level may not be justified if they lead to large adverse changes in social costs at the local level. Recognizing these problems, EPA is developing an approach for estimating national and local costs and benefits (regulatory impact analysis) that in support of regulatory decision making. Information collected from these efforts will be used to support another negotiated rule making process, scheduled to begin at the end of 1998.

References

1. U.S. EPA. National Interim Primary Drinking Water Regulations; Control of Trihalomethanes in Drinking Water. *Fed. Reg.* 44:231: 68624-68707. November 29, 1979.
2. WHO. Guidelines for Drinking Water Quality. Geneva. 1993
3. U.S. EPA. National Primary Drinking Water Regulations; Disinfectants and Disinfection Byproducts; Proposed Rule. *Fed. Reg.*, 59:145:38668-38829. July 29, 1994
4. U.S. EPA. National Primary Drinking Water Regulations; Enhanced Surface Water Treatment Requirements; Proposed Rule. *Fed. Reg.*, 59:145:38832. July 29, 1994
5. U.S. EPA, National Primary Drinking Water Regulations; Disinfectants and Disinfection Byproducts; Proposed Rule. *Fed. Reg.*, 62: 212: 59388-59484. November 3, 1997.
6. U.S. EPA, National Primary Drinking Water Regulations; Interim enhanced Surface Water Treatment Rule Notice of Data Availability; Proposed Rule. *Fed. Reg.*, 62:212: 59486-59557. November 3, 1997.
7. U.S. EPA, DBP Stage 2 Rule and Long-Term 2 Enhanced Surface Water Treatment Rule: Draft Report on Research to Support Rules. Office of Ground Water and Drinking Water. USEPA. November 12, 1997.

DBP STANDARDS AND REGULATION: THE UK AND EUROPEAN SITUATION

Owen Hydes

Deputy Chief Inspector
Drinking Water Inspectorate
Ashdown House
123 Victoria Street
London SW1E 6DE

1. SUMMARY

This overview of the UK and European situation on disinfection byproduct (DBP) standards and regulations has been prepared in part from a questionnaire sent to the regulators in the 15 Member States of the European Union. It describes the relevant requirements of the current European Community Drinking Water Directive and the way in which they have been implemented by 12 of the Member States. It shows the wide variations in disinfection practice and DBP standards in these countries.

It sets out the guideline values for DBPs contained in the World Health Organization's 1993 Guidelines for Drinking Water Quality and the addendum for the rolling revision of those guidelines. Finally it outlines the proposals for standards and monitoring for DBPs in the new European Community Directive and discusses the challenges that these proposals, if agreed, will present to the regulators.

2. EUROPEAN COMMUNITY DRINKING WATER DIRECTIVE

2.1 General

The Member States of the European Community are required to meet the standards and other requirements of the EC Directive relating to the quality of water intended for human consumption (80/778/EEC)[1] which was agreed in 1980 and came into effect in 1985. The Directive sets standards for various microbiological, physical and chemical parameters and prescribes minimum monitoring frequencies for key parameters. It is necessary to consider the Directive's requirements for disinfection because these are relevant to its requirements for DBPs.

2.2 Disinfection and DBPs

The Directive does not specifically require water supplies to be disinfected. However, it sets mandatory standards for microbiological parameters, requires frequent monitoring   for coliforms and faecal coliforms and requires twice the monitoring frequency for supplies which must be disinfected. Also it requires frequent monitoring for residual chlorine or other disinfectants when disinfection is practised. So although there is no specific requirement for disinfection there is a strong implication that disinfection of some or many water supplies is expected.

The Directive does not set standards for DBPs. It contains a parameter called organochlorine compounds not included in the pesticide parameter and has a non-mandatory guide level of $1\mu gl^{-1}$ for these substances, but no mandatory standard. However, there is a comment that haloform concentrations should be as low as possible.

2.3 Implementation of the Directive in Member States

All Member States have implemented the mandatory standards for microbiological parameters. Many Member States have gone much further than required by the Directive by including in their national legislation requirements for disinfection and standards for DBPs.

Table 1 shows which of 12 Member States have a national legal requirement to disinfect water supplies. Three countries require all supplies to be disinfected, and four more require surface water derived supplies to be disinfected, but not groundwater derived supplies. Most of the countries that do not legally require disinfection have issued guidance to their water suppliers on the situations where they are expected to disinfect water supplies, particularly to ensure that the microbiological standards are met. Some countries specifically state that disinfection is not required for good quality groundwaters which are

Table 1 *Legal requirement to disinfect*

Country	Surface Water Supplies	Groundwater Supplies
Austria	Yes	No
Belgium	No	No
Denmark	Yes	No
Finland	No	No
France	Yes	No
Germany	No	No
Ireland	No	No
Netherlands	Yes	No
Portugal	Yes	Yes
Spain	Yes	Yes
Sweden	No	No
United Kingdom	Yes	Yes

microbiologically satisfactory and for some surface waters which are either of very good microbiological quality or have sufficient other treatment barriers to ensure that micro-organisms are removed without the need for disinfection.

Of the three countries that legally require all supplies to be disinfected, Spain and Portugal also legally require a disinfectant residual to be maintained in distribution. Germany and Austria also require a disinfectant residual but only when the use of a disinfectant is necessary to achieve microbiological standards. A number of other countries have issued guidance to water suppliers on the disinfectant residual that should be aimed for in distribution. Table 2 shows the legal requirements and the guideline values for various disinfectants in these countries.

Table 2 - *legal disinfectant residuals (S) and guideline values (G) (mgl^{-1})*

Country	Disinfectant	Leaving treatment works		Supplied to user	
Austria	chlorine (free)	0.3 - 0.5	(S)	<0.3	(S)
	ozone	> 0.1	(S)	<0.05	(S)
	chlorine dioxide	>0.05	(S)	<0.2	(S)
Germany	chlorine (free)	-		>0.1	(S)
	chlorine dioxide			0.05 - 0.2	(S)
Portugal	chlorine	-		0.2	(S)
Spain	chlorine (free)	-		0.2 - 0.8	(S)
	chlorine (combined)	-		1.0 - 1.8	(S)
Belgium	chlorine (free)	0.2	(G)	-	
Finland	chlorine (total)	<1.0	(G)	-	
France	chlorine (free)	0.1	(G)		
Ireland	chlorine (free)	0.2 - 0.5	(G)		

Although the Directive does not require standards to be set for DBPs, each of the 12 Member States except France, Portugal and Spain have set standards for one or more DBPs. Some countries also have guidelines for some DBPs. Unless otherwise stated, these standards or guidelines apply at the point at which the water supplier makes the water available to the user. Table 3 summarises the standards set or guidelines given. The main group of DBPs for which standards are set are the trihalomethanes with values ranging from 10 to 100μgl^{-1} for those countries which have standards for total substances and up to 200μgl^{-1} for those countries which have standards for individual substances. It is noticeable and not surprising that the more stringent standards have been set by some of the countries that do not require disinfection of water supplies. A few countries have a standard for chlorite/chlorine dioxide/chlorate ranging from 200 to 500μgl^{-1}. Only one country, the Netherlands, has a guideline for bromate.

Table 3   *DBP standards/guidelines*

Country	Disinfection byproduct	Standard (μgl^{-1})
Austria	Trihalomethanes (total chlorite	30 200
Belgium	Trihalomethanes (total)	100
Denmark	Trihalomethanes (total)	as low as possible 10 - 15 (guideline)
Finland	Chloroform Bromodichloromethane Others	200 60 advise WHO guideline
Germany	Trihalomethanes (total)(1) Chlorite/chlorine dioxide	10 (ex works) 25 (special circumstances) 200 (ex works)
Ireland	Trihalomethanes (total)	100
Netherlands	Halogenated hydrocarbons Bromate Trihalomethanes (total)	1 (each substance) Guidelines 0.5 5 (when disinfection used) 20
Sweden	Trihalomethanes (total) Trihalomethanes (total)	50 2 (guideline)
United Kingdom	Trihalomethanes (total) Chlorite/chlorine dioxide	100 (average) 500

The Netherlands also has some provisional guideline values for DBPs which are shown in Table 4.

Table 4   *Netherlands provisional guideline values*

Disinfection byproduct	Provisional guidelines (μgl^{-1})
chloroform	5
bromodichloromethane	6
dibromochloromethane	5
bromoform	5
chlorite	200
trichloroacetic acid	100
cyanogen chloride	70
dibromoacetonitrile	100
dichloroacetonitrile	90
dichloroacetic acid	50
formaldehyde	90
trichloroacetonitrile	1
chloral hydrate	10

3. WORLD HEALTH ORGANIZATION GUIDELINES

3.1 Balancing microbiological quality and DBPs

The 1993 Guidelines for Drinking Water Quality[2] indicate that the provision of a safe supply of drinking water from a microbiological viewpoint depends on the use of either a protected high quality groundwater or a properly selected and operated series of multibarrier treatment processes, including terminal disinfection, capable of removal or inactivation of harmful microorganisms. The Guidelines recognise and stress that disinfection is unquestionably the most important step in the treatment of water supplies. The paramount importance of disinfection and microbiological quality requires some flexibility in the derivation and application of guideline values for disinfectant residuals and DBPs. This is possible because of the large margin of safety incorporated into the guideline values for these substances. The Guidelines specifically indicate that when a choice has to be made between meeting microbiological guidelines and guidelines for DBPs, microbiological quality must take precedence and efficient disinfection must never be compromised.

3.2 Guidelines for DBPs

For those DBPs that are regarded as potential genotoxic carcinogens the guideline values are determined from a mathematical model and the values given are the concentrations   in drinking water associated with an estimated excess lifetime cancer risk of 10^{-5}, that is one additional case of cancer per 100,000 population ingesting drinking water containing the substance at the guideline value for 70 years. For other DBPs that are regarded as non-carcinogenic or non-genotoxic carcinogens the guideline values are determined from a tolerable daily intake (TDI) derived from appropriate toxicity studies as follows:

$$TDI = \frac{NOAEL \ or \ LOAEL}{UF}$$

where NOAEL = no observed adverse effect level
LOAEL = lowest observed adverse effect level
UF = uncertainty factor obtained by expert judgement of the nature and quality of the toxicological information.

$$Guideline \ value = \frac{TDI \times BW \times P}{C}$$

where BW = body weight (60kg for adults)
P = fraction of TDI allocated to drinking water
C = daily drinking water consumption (2 litres for adults)

Table 5 shows the guideline values derived by WHO. Provisional guideline values are given when:

(1)   the toxicity data available is limited such that the uncertainty factor is greater than 1000; or

(2)   the calculated value is below the detection limit of the analytical method (the provisional value is related to the detection limit); or
(3)   the calculated value is below what can be achieved by practical treatment methods (the provisional value is what can be achieved by treatment); or
4)   disinfection may result in guideline value being exceeded.

Table 5 *Guidelines for DBPs*

Disinfection byproduct	Guideline value (μgl^{-1})	Comments
Bromate	25 (P)	Detection limit of analytical method, 7×10^{-5} excess cancer risk.
Chlorite	200(P)	Limited toxicity data.
2,4,6 -Trichlorophenol	200	10^{-5} excess cancer risk. Taste or odour of water may be affected at lower concentrations
Formaldehyde	900	
Bromoform	100	
Bromodichloromethane	60	10^{-5} excess cancer risk.
Dibromochloromethane	100	
Chloroform	200	10^{-5} excess cancer risk
Dichloroacetic acid	50 (P)	May be exceeded with disinfection.
Trichloroacetic acid	100 (P)	Limited toxicity data. Maybe exceeded with disinfection.
Chloral hydrate	10 (P)	Limited toxicity data.
Dichloroacetonitrile	90 (P)	Limited toxicity data.
Dibromoacetonitrile	100 (P)	Limited toxicity data.
Trichloroacetonitrile	1 (P)	Limited toxicity data.
Cyanogen chloride (as CN)	70	

WHO does not give a value for total trihalomethanes but it suggests a fractionation approach to account for additive toxicity for authorities wishing to establish a total trihalomethane standard. The sum of the ratio of the concentration of each individual trihalomethane to its respective guideline value should not exceed 1. Application of this rule in many circumstances leads to a guideline value approximating to 100μgl^{-1} for total trihalomethanes.

3.3 Review of WHO Guidelines

In 1995 WHO commenced a rolling review of its 1993 Guidelines. The initial review considered 20 parameters including chloroform. The results of this review have been published in an addendum to the guidelines[3]. On chloroform the review concluded that the weight of evidence is that it is not a genotoxic carcinogen. Therefore the guideline value should be calculated from a TDI derived from its chemical toxicity. On this basis the guideline value is $200\mu gl^{-1}$, which is the same value as originally derived on the basis of potential genotoxicity with a 10^{-5} excess cancer risk.

4. REVISED EUROPEAN COMMUNITY DIRECTIVE

4.1 Commission's proposals

The Commission of the European Communities Directive published its proposals for a new drinking water directive in January 1995[4]. The 1993 WHO guidelines were taken into account in preparing the proposals. Table 6 shows the proposed standards for DBPs. The Commission's Explanatory Memorandum indicates that bromate is included because it is carcinogenic and may be formed by oxidation of bromide present in raw water or introduced by some treatment methods. It concluded that the lowest value consistent with reliable disinfection with chlorine is $10\mu gl^{-1}$ and this can be measured with conventional techniques of analysis.

On trihalomethanes the Explanatory Memorandum indicates that bromodichloromethane and chloroform are included because they are carcinogenic. The Commission's general view is that standards for carcinogens should be based on an excess lifetime cancer risk of 10^{-6} which would correspond to $6\mu gl^{-1}$ for bromodichloromethane and $20\mu gl^{-1}$ for chloroform. However, it recognised that in practice it can be difficult to achieve those levels and therefore it proposed values of 15 and $40\mu gl^{-1}$ respectively at the outlet of the treatment plant. The Commission indicated that it would study the extent to which the concentrations of these compounds increases during distribution.

Table 6 *Proposed values for DBPs*

Disinfection byproduct	Proposed standard μgl^{-1}	Comments
Bromate	$10\mu g/l$	consistent with disinfection
Bromodichloromethane	$15\mu g/l$	}at outlet of works.
Chloroform	$40\mu g/l$	}achievable in practice.

4.2 Commission's study of trihalomethanes

The European Commission contracted the Joint Research Centre, Ispra-Environment Institute to carry out a study to investigate the formation of trihalomethanes during treatment and in the distribution network and the practical possibilities for reducing concentrations of these compounds without compromising disinfection. Two reports of this study have been published[5,6]. The main conclusions included:

(1) effective disinfection is paramount. Efforts to reduce concentrations of byproducts from all disinfectants should never compromise effective disinfection;

(2) due to the wide variation in standards and lack of standards and the wide variation in monitoring strategies and lack of monitoring within Member States, it was not possible to provide quantitative estimates of trends in population exposure to trihalomethanes;

(3)   the proposed standard of $55\mu g l^{-1}$ for combined chloroform and bromodichloromethane ($40/15\mu g l^{-1}$ respectively) is technically achievable. It is also technically feasible to reduce these concentrations to $20/6\mu g l^{-1}$ respectively (10^{-6} excess cancer risk). But it noted that Member States which rely heavily on unfiltered surface water for their supplies will require changes to treatment processes, considerable investment and a reasonable time scale to achieve compliance with the proposed standard; and

(4)   the distribution system plays an important role in the final trihalomethane concentration in the water delivered to the consumer. Many factors affect the changes in trihalomethane concentrations from treatment works to the consumers' taps. The study found from a few well documented case studies that trihalomethane concentrations increased between 1.2 and 3 times between the treatment works and consumers' taps and the increase was typically 2 times when the residence time was less than 48 hours.

4.3 European Parliament

The European Parliament in its first reading of the proposed new Directive adopted an amendment of $80\mu g l^{-1}$ at consumers' taps for the sum of the concentrations for the 4 trihalomethanes, chloroform, bromodichloromethane, dibromochloromethane and bromoform.

4.4 Member States Common Position

The negotiations on the proposed new Directive in the Environment Working Group and in the Council of Environment Ministers took account of the results of the European Commission's study and the European Parliament's proposed amendment.

The common position agreed by the Council of Environment Ministers in late 1997 retained the proposed $10\mu g l^{-1}$ standard for bromate to be met within 10 years with a $25\mu g l^{-1}$ interim standard to be met within 5 years and a comment that, where possible, Member States should strive for a lower value without compromising disinfection. However, it modified the proposal for trihalomethanes by agreeing a total trihalomethane standard of $100\mu g l^{-1}$ at consumers' taps to be met within 10 years with an interim standard of $150\mu g l^{-1}$ to be met within 5 years. It also included comments that, where possible, Member States should strive for a lower value without compromising disinfection and that appropriate measures should be taken to reduce concentrations as much as possible during the period needed to achieve compliance.

The proposed Directive also requires monitoring for DBPs at the audit monitoring frequencies. These are minimum frequencies which depend on the volume of water supplied and range from one per year for supplies of 100 m^3/day to 4 per year for supplies of 1000 m^3/day to over 15 per year for supplies of over 100,000 m^3/day.

The common position is currently being considered by the European Parliament in its second reading of the proposed new Directive. Any amendments proposed by the Parliament will need to be considered by the European Commission and the Council of Environment Ministers. It is expected that the new Directive will be finally agreed in Autumn 1998.

5. CHALLENGES FOR THE REGULATORS

The regulators in the Member States will be required to include in their national legislation the mandatory standards for DBPs and the minimum monitoring frequencies.

The challenges for these regulators will be:

(1)   whether to set tighter standards for bromate or total trihalomethanes. Some countries already have tighter standards for total trihalomethanes;

(2)   whether to set standards for individual trihalomethanes. Some countries already have such standards;

(3)   whether to set standards for other DBPs, for example those included in the 1993 WHO Guidelines; and

(4)   whether to set higher monitoring frequencies for DBPs, particularly for the smaller surface water derived supplies where the minimum monitoring frequency may not be adequate to determine compliance throughout a year.

It is expected that there will be variations in the standards adopted by Member States with some opting for more stringent standards or standards for additional DBPs.

References

1.   Official Journal of European Communities, No. 229, 30 August 1980.
2.   World Health Organization, Guidelines for Drinking Water Quality, Second Edition, Volume 1, Geneva 1993.
3.   World Health Organization, Guidelines for Drinking Water Quality, Second Edition, Volume 2 Addendum, Geneva 1998.
4.   Commission of the European Communities, COM(94) final, Brussels 1995.
5.   Environment Institute, Standards and Strategies in the European Union to Control Trihalomethanes in Drinking Water, 1997.
6.   Environment Institute, Exposure of the European Population to Trihalomethanes in Drinking Water, Volume 2, 1997.

MEETING STRICT DBP STANDARDS - THE VIEW OF THE EUROPEAN WATER INDUSTRY

Dr R A Breach

Head of Quality & Environmental Services,
Severn Trent Water

Chairman, EUREAU Commission 1

1   INTRODUCTION

As part of the process of formulating a new EU Drinking Water Directive, there has been extensive discussion on the issue of disinfection byproduct (DBP) standards. Water in Europe is characterised by a wide diversity of water resource types, and supply/treatment infrastructure.   This has a significant impact on the level and type of disinfection byproducts found in different supplies around Europe.   Therefore, in trying to derive common but practicable European DBP standards this diversity must be taken into account in a way that achieves cost effective but high levels of health protection.

In Europe, as elsewhere, there has always been good collaboration between water suppliers themselves, as well as research agencies and regulatory bodies.   This collaboration has been amply demonstrated during the process of developing the revised Drinking Water Directive.   EUREAU, the association of European Water Suppliers and Wastewater Operators, has been actively involved in the discussion on the new Directive, focusing particularly on provision of practical advice and guidance. This involvement has been publicly welcomed by the European Commission.

This paper describes some of the issues relating to the achievement of strict DBP standards from the viewpoint of water suppliers and their customers.   The new EU Directive is likely to be adopted by the end of 1998, and will substantially update and improve the regulatory framework for drinking water quality across all European Union Member States.

2   THE NEW DRINKING WATER DIRECTIVE

The original Drinking Water Directive was adopted in 1980 to come into force in 1985. During the late 1970s, when the content of the existing Directive was being discussed, knowledge and concern about disinfection byproducts was relatively low. For that reason, no specific standards were set for DBPs in the Directive, although there was a poorly defined "guide value" for organochlorine compounds, which made reference to the fact that disinfection byproducts should be kept as low as possible. As knowledge of DBPs in drinking water increased, Member States tended to adopt different DBP standards, which reflected the situation in their own country. There is thus at present no common approach to control of DBPs across the European Union.

Discussions about updating the Drinking Water Directive in the light of practical experience as well as new scientific knowledge took place during the period 1992 to 1995, with a draft proposal first published by the European Commission in 1995. This recognised that there should be a move towards common European standards for DBPs, but acknowledged that at that time there was insufficient information on the current situation on DBPs in different Member States to make a final decision. Provisional DBP standards were therefore set based on values leaving source works, but a major study of trihalomethanes in Europe was commissioned from the Joint Research Centre (JRC) of the European Commission at Ispra in Italy. This comprehensive study[1,2] provided an excellent base of practical information on trihalomethanes and related DBPs.

Following further extensive discussion and consultation, the final version of the Directive [3] is likely to include standards of 150 ug/1 total THMs at the tap to be achieved in 5 years, and 100 ug/1 total THMs at the tap within 10 years. A standard is also proposed for bromate at 25 ug/1 to be achieved in 5 years, and 10 ug/1 in 10 years.

One of the fundamental principles in deriving DBP standards, which was universally accepted by all parties, was the fact that disinfection effectiveness is paramount. At no time should the understandable desire to minimise disinfection byproduct levels ever compromise the ability of water suppliers to secure adequate disinfection of water. EUREAU totally supported this principle and believes that the proposed standards in the Directive do protect public health, whilst at the same time providing no unreasonable constraint to effective disinfection.

3   DISINFECTION PRACTICE IN EUROPE

The report by JRC provided a useful summary of the current disinfection practices in different Member States. This is summarised in Table 1. In general, chlorine is still the dominant disinfectant in most countries, although there are moves to reduce or even phase out chlorine disinfection in a number of countries, particularly the Netherlands, Germany and Denmark. As well as chlorine, other commonly used disinfectants include ozone, ultraviolet light, and chlorine dioxide. There is however much greater variation between countries in the extent to which they use these latter three disinfectants, with, for example, chlorine dioxide being extensively used in Germany and Italy, and to a lesser extent France and the Netherlands, but hardly at all in some other countries. Chloramination is not used as a primary disinfectant, but is used as a network disinfectant in a small number of countries.

Table 1 *Disinfection Practices in European Countries (references 1 & 2)*

	At	Be	Dk	De	Es	Fi	Fr	GB	Gr	Ir	It	NL	Pt	Sw
Cl2	3	3	↑	3	3	3	2	3	3	3	3	2	3	3
O3	1	1		2	2	1	2	1			1	3		
UV	1	1		1		1		1				1		
ClO2	1	1		3	1	1	2	1			3	2		1
Clm					2	1		1						2

1 = occasional 2 = common 3 = dominant
Clm = chloramination

4   PRACTICAL ISSUES IN SETTING DBP STANDARDS

During the discussion about DBP standards, a number of important practical issues were identified. These included the following:

4.1   The Variable Nature of the Water Resource and Supply Infrastructure

The balance between surface and groundwater resources in different European countries varies widely. In some Member States virtually all of the water is obtained from groundwater, whereas other countries rely extensively on surface water. In general, there is likely to be a nett shift in European water resource use from groundwater to surface water, because of both the diminishing availability of sustainable groundwater reserves, and the potential accession of new Member States which have a higher available proportion of surface water resources.   Inevitably surface waters tend to have higher levels of natural organic DBP precursors and microbiological contamination than groundwater and it is therefore much more difficult to reliably achieve very low levels of DBPs with such resources.

4.2   Treatment Flexibility

One particularly important issue is that unlike many parameters, it is not possible to measure disinfection byproducts by on-line monitoring at water treatment works. Such parameters can only be analysed in a laboratory. It is therefore very difficult to introduce real time process control, as is the case with other parameters such as turbidity and coagulant residual.

The level of DBPs is potentially closely correlated with the level of precursors in the raw water. In many surface water plants, climatic and other factors mean that these precursors can vary significantly, often over a short period of time. Without on-line monitoring, it is therefore impossible to know the exact level of disinfection byproducts created at the treatment plant at any point in time. Because of the paramount priority to achieve effective disinfection, it is normal practice to use sufficient chemical disinfectant to suppress any residual short tenn oxidant demand, and achieve predetermined CT values in order to ensure pathogen kill. For this reason it is difficult to reliably control DBP levels by day to day operation of the plant, particularly if raw water precursors are variable. Such control of DBPs can only to be built into the fundamental process design of the plant, for example by reducing DBP precursors prior to addition of a chemical disinfectant.

4.3   Which Particular DBP to Control?

Research has identified many different potential byproducts that can be formed, albeit often at extremely low levels, and well below any concentration posing a risk to health. Analysis of many of the more exotic DBPs is complex and expensive, particularly for smaller laboratories.   Although there is often no simple or direct correlation between different DBP species, as a general rule the overall level of many DBPs tends to increase broadly at the same rate. In regulatory terms, it is therefore quite legitimate to consider the concept of indicator DBPs.   These are substances which can be relatively easily measured and provide a general indication of the overall level of certain groups of DBPS, without the need for complex and expensive analysis of all such compounds. In the new European Directive, total trihalomethanes has been identified as the best overall indicator for chlorinated organic DBPs, with, in addition, a specific standard for bromate in view of the particular concern about its potential toxicity.   There are of course other possible indicators for organic DBPs, for example total haloacetic acids, but standards have not been currently set for them in the European Drinking Water Directive.

4.4 Balancing Risks

In setting DBP standards it is particularly important to take a holistic approach to assessing the risks from such substances. All chemical disinfectants to some extent produce byproducts, each of which must be looked at individually. However the type of disinfectant used and the treatment process will inevitably produce a different range of DBPs. For example, changes to treatment to reduce chlorinated haloforms can actually increase the level of brominated DBPs. Given that the toxicity of brominated compounds is potentially of greater concern, the question has to be asked as to whether it is better in health terms to have slightly more relaxed standards for chlorinated compounds thus allowing brominated compounds to be kept as low as possible. Similar arguments relate to the distinction between halogenated byproducts produced by chlorine based disinfection, and oxygenated byproducts produced by ozone disinfection, or between organic DBPs and inorganic DBPs such as bromate, chlorate and chlorite. Table 2, set out below, demonstrates that there is no perfect chemical disinfectant from the point of view of disinfection byproducts. Equally however, poor disinfection is much worse in risk to human health. In setting DBP standards therefore, an appropriate balance has to be struck in setting standards for the different DBPs which provides an overall good level of health protection, whilst allowing water suppliers the flexibility to choose the appropriate disinfectant for different circumstances. It would be foolish indeed to set such strict standards for a range of DBPs, that the use of any chemical disinfectant is effectively precluded.

Table 2 *Chemical Disinfectants and Main DBPs*

Gaseous chlorine	THM & Other halo organics
Hypochiorite	THM & Other halo organics
	Bromate
	Chlorate
Chlorine dioxide	Chlorite
	Chlorate
Ozone	oxy organics
	Bromate
Chloramination	Nitrite

4.5 Compliance Interpretation

A final practical issue in setting DBP standards is the way in which compliance is interpreted. As with a number of parameters, the statistical distribution of DBP results tends to be log normal, and skewed to upper values. This means for example that there is a considerable difference in practical impact between a DBP standard set as a maximum, or an average. For example the proposed long term total trihalomethane standard in the European Drinking Water Directive is 100 ug/1 as a maximum. This is probably more stringent than the equivalent US EPA standard of 80 ug/1 as an average, even though numerically the US standard superficially might appear to be tighter.

It is also very important to define where the standard applies. Historically in some countries, the DBP standard has been set at the outlet from the treatment plant.

However, research has shown that the level of DBPs can increase significantly in the distribution network, particularly if this is long and reticulated. This occurs because reaction continues to take place between disinfectant residuals necessary for bacteriological security and DBP precursors. The ratio of tap to plant DBP values varies considerably, but typically can be a factor of 2. In the European Directive the DBP standard has been set, rightly, to apply at the customer's tap, since this is the water that is consumed. However, by setting a level at the tap, implicitly the level leaving the treatment plant must be much lower to allow sufficient leeway for increases within the distribution network.

5   A CUSTOMER PERSPECTIVE

The debate about disinfection byproducts tends to be very technical, and take place mainly between experts, whether from water suppliers, research agencies, or regulatory bodies. However, as in all things, it is also important to look at the question of DBP standards from a customer viewpoint. Priorities that would be set by a customer are:

(i)   that the water must be "safe"
(ii)   that it must taste good
(iii)   that it must be affordable.

The question therefore emerges, "how safe is safe enough?", when looking at the risks of disinfection byproducts. The risk from microbiological contamination is relatively well understood, and of much greater concern than that from DBPs. The assessment of risk from DBPs is still subject to some uncertainty, despite massive research effort over the last 20 years. However a growing consensus seems to be emerging about the levels of DBPs below which there are unlikely to be any significant or measurable risks.

The relationship between the cost difficulty and the stringency of DBP standards is exponential. The consequence is that whilst standards of 100 ug/1 total THMS, and, say, half that level, are probably toxicologically indistinguishable, or at least within the normal range of scientific uncertainty, the practical and cost impact of achieving the tighter standard could be very severe. Better technology can minimise but not eliminate that massive cost increase. It is thus quite legitimate to ask whether the health benefit derived from continued further reduction in DBP levels is justified by the cost. That is of course a question for regulators, not for water utilities, but does need to be addressed.

However, customers do want water that tastes good. Often the main source of complaints about taste is chlorine levels. Therefore water suppliers have to optimise disinfection both to achieve acceptable taste levels and to minimise DBP levels, whilst at the same time maintaining high levels of disinfection security. There is an argument that water which meets customer requirements in terms of disinfectant taste, may already have acceptably low levels of DBPs in terms of health risk. The extent to which even lower levels of DBPs should be achieved is a difficult matter of judgement for regulators.

Intrinsically, the ease of achieving quality targets and customer requirements is much higher in good quality local groundwater systems than with poorer quality raw surface water, particularly when associated with long distribution networks. It is of course possible to achieve high quality with surface derived systems but usually only with significant expenditure and effort. Since the proportion of surface water in Europe is likely to increase, it is therefore imperative that European DBP standards are set at values which not only protect health but can be cost effectively achieved in surface water as well as groundwater.

6 SUMMARY

The establishment of European DBP standards must balance a number of complex factors. Critical amongst these is the need to ensure effective disinfection, whilst keeping DBPs at the lowest practicable level. Standards must be based on best scientific advice, but not be set at levels that can only be achieved with groundwater sources, unless disproportionate costs are incurred.

Water suppliers in Europe believe that the approach adopted in the new European Drinking Water Directive does achieve a sound compromise between these potentially conflicting objectives, and allows water to be produced which provides high levels of health protection but at affordable prices.

References

1   Standards and Strategies in the European Union to Control
    Trihalomethanes(THMS) in Drinking Water, 1997.
    EC Joint Research Centre (JRC - Ispra).

2   Exposure of the European Population to Trihalomethanes (THMS) in Drinking
    Water, Volume 2,1997.
    EC Joint Research Centre (JRC - Ispra).

3   Common Position of the Council (1 3.98) with a view to adoption of Council
    Directive on the Quality of Water for Human Consumption (OJ.C91.1, 26.3.98).

NORTH WEST WATER'S OPTIONS FOR COMPLIANCE WITH THE PROPOSED REVISION OF THE EUROPEAN DRINKING WATER DIRECTIVE STANDARD FOR DBPS.

Frank White

Quality Regulation Manager
Water Management
North West Water
Thirlmere House
Warrington WA5 3LP

1 INTRODUCTION

North West Water (NWW) supplies drinking water to 6.8 million people in the North West of England. NWW serves an area of 14,000 square kilometres using 173 water treatment works, 960 km of aqueducts, 396 service reservoirs and 40,000 km of distribution mains. Typical daily output is 2,200 megalitres, of which 67% comes from upland storage reservoirs 25%, from river sources and 8% from boreholes.

Aqueducts are widely used by NWW as an integrated and cost effective way to convey water, taking advantage of gravity in order to reduce pumping costs. Major aqueducts transport water over considerable distances to the large conurbations from the rural upland reservoirs and lakes.[1]

Much of the region's raw water is soft, coloured, poorly buffered and contains dissolved metals, such as iron and manganese. In addition, much of the older housing stock in the region contains lead service connections, which can leach lead into drinking water. These factors provide a significant challenge to provide a stable, wholesome, aesthetically pleasing and reasonably priced product that meets the expectations of customers.

Disinfection of water is the primary concern to protect the public from water borne pathogens. The application of oxidants such as chlorine and ozone to water can produce disinfection by-products (DBPs). The proposed European Union Directive concerning the quality of water intended for human consumption (subsequently referred to as the new Directive) contains standards/Parametric Values (PV) for some of these DBPs.[2] NWW is reviewing its options for securing or facilitating compliance with those standards whilst providing a wholesome product at a reasonable price.

2 PROPOSED NEW STANDARDS FOR DBPs

The proposed PV for Total Trihalomethanes (THMs) in the new Directive is 100μg/l. The specified THMs are chloroform, bromoform, dibromochloromethane and bromodichloromethane. This PV must be complied with within 10 years of the Directive coming into force. A standard of 150μg/l is permitted for the first 5 years.

The new Directive also sets a PV for Bromate at 10μg/l. This PV must be complied within 10 years. A standard of 25μg/l is permitted for the first 5 years.

The specified THMs are the same parameters as are currently measured by NWW to comply with the Water Supply (Water Quality) Regulations (1989) as amended [3] (subsequently referred to as The Regulations). The Regulations measure total THMs as a three month rolling average with a Prescribed Concentration or Value (PCV) of 100μg/l.

We anticipate the standard of 100μg/l will come into force immediately the Regulations are transposed into UK law.

Within NWW THM data have been collected from customer's taps for regulatory purposes, but very little data are available from service reservoirs or water treatment works.

Currently no PCV exists for bromate and little or no data has been collected.

3 FACTORS INFLUENCING THE FORMATION OF THMS AND BROMATE

3.1 THMs formation

THM formation is promoted by alkaline pH, long disinfection chlorine contact times and the presence of naturally occurring organic matter e.g., humic and fulvic acids.

3.1.1 Alkaline pH: Although most of NWW supplies are naturally acidic, the pH is raised by lime or caustic soda, in order to reduce its corrosivity and plumbosolvency, typically with ex-works of pH 8.5. Further pH rise may occur in cement mortar lined mains as the waters are poorly buffered.

3.1.2 Long disinfection contact times: THM formation is dependant on the time of contact and the concentration of chlorine. Total chlorine concentration in water leaving the works has to be high (often in excess of 2 mg/l) because of the high chlorine demand of the treated waters and the need to have high chlorine doses to achieve adequate disinfection at high pH values. The long length of the aqueducts and the need to use secondary disinfection from takeoffs and also within the reticulation substantially lengthens total contact times.

3.1.3 Temperatures: Warmer temperatures favour THM formation, suggesting an increase with water temperature in summer months (although this may be mitigated by decreased contact time as a result of higher demand on the network)

3.1.4 Humic and fulvic acids: The catchments of most upland sources are from exposed peat moors overlying hard granitic rocks. The run off is thus rich in humic and fulvic acids and loading varies seasonally.

3.2 Bromate

Commercially available hypochlorite does contain concentrations of Bromate which may cause NWW difficulty in compliance with the PV. This is used at many NWW sites but alternative disinfection systems exist such as chlorine gas and onsite electrolytic chlorination.

Bromate formation can occur as a by-product of Ozonation or electrolytic chlorination. In both cases bromide is oxidised to Bromate. Bromate formation is favoured at alkaline pH, high ozone dose and high bromide ion concentration.

Ozone use is restricted to one site in NWW, for colour removal and not disinfection.

Electrolytic chlorination is practised widely in NWW, but the salt used to generate brine is low in bromide and not thought to be a major problem.

4 COMPLIANCE WITH EXISTING DBP REGULATORY STANDARDS (1997)

In 1997, only nine of 306 NWW water supply zones failed to meet the three month rolling average standard for THMs [3]. If the standard were an absolute value, 55 of our 306 water supply zones would fail to meet the standard. No regulatory data exists for Bromate compliance [2].

To comply with the new standards for DBP will require adjustment to the operation of many of our treatment regimes and in some cases the building of new unit processes.

The key sources are described below:

4.1 Upland sources

The catchments where the water is collected are used for upland grazing, forestry, and have peaty soils overlying hard granitic rocks. Water collected in upland reservoirs has to be moved to the lowland areas of the region where the majority of users are situated. NWW uses 960 km of aqueducts to conduct water to its customers. The system has developed over 150 years and is highly interlinked and flexible in its operation[1].

The raw water is typically poorly buffered acidic to neutral pH and contains humic and fulvic acids causing the water to be coloured. Dissolved metals are also present: namely iron, manganese and aluminium. (Table 1).

The three longest aqueduct systems are the Thirlmere aqueduct (TA), the Haweswater aqueduct (HA) and the Vyrnwy aqueduct (VA) see Figure 1.

4.1.1 Thirlmere system: The TA system is 154km long and was commissioned in 1894. It comprises a series of single or multi pipe siphons interconnected by single line tunnels. It conveys 220Ml/d from the Lake District to parts of south Cumbria, Lancaster, the Fylde coast, Preston, Wigan, Bolton and Greater Manchester. Residence time in the system can be up to seven days. The aqueduct terminates at Lostock storage reservoir and enters the Manchester ringmain after pH adjustment and secondary chlorination.

The TA is supplied principally by Thirlmere reservoir. The water entering the aqueduct is treated at Dunmail Raise water treatment works. It undergoes coarse screening, microstraining, disinfection and pH correction with caustic soda .En route to Manchester ringmain there are numerous takeoffs ranging in size from 0.01-100Ml/d. The TA supply can be further supplemented by a treated water supply from Haweswater treatment works.

The TA was built at a time when water quality standards were less exacting than today. The construction methods used at that time leave it potentially vulnerable to ingress of ground water with the attendant risks of contamination. All the takeoffs undergo further secondary disinfection as a result. This allows further contact time for the formation of THMs.

4.1.2 Haweswater system: The aqueduct is 117km long and ranges in age from 20-57 years old. It supplies parts of south Cumbria, the Fylde coast, Blackburn, Burnley Rossendale and greater Manchester.

Figure 1 North west water catchment showing upland sources, key surface water sources

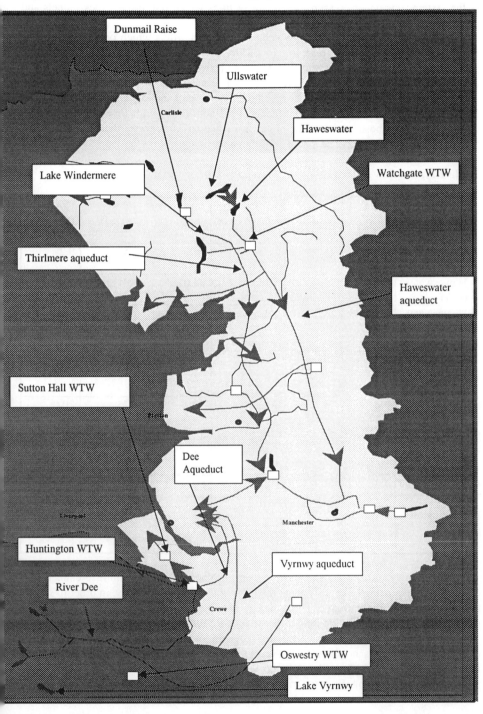

It is a more modern construction than the TA and is less subject to groundwater ingress as the system runs under pressure. There are sections which run through rock tunnels and here ingress of water is a possibility.

Haweswater aqueduct is fed by Watchgate treatment works which obtains most of its raw water from Haweswater reservoir (500Ml/d is typical). Wet Sleddale reservoir and pumped supply from Lake Ullswater can feed into Haweswater to supplement the supply to the reservoir. A pumped supply from Lake Windermere can also supply the treatment works. At Watchgate WTW, raw water is filtered using sand anthracite rapid gravity filters. The water is then dosed with chlorine gas and pH adjustment with lime is made to help minimise the plumbosolvency of the water. En route to Manchester there are numerous takeoffs, all of which receive secondary chlorination.

Table 1 *Typical Raw water quality for key upland sources and river Dee sources*

Source	PH	Hardness (mg/l as CaCO3)	Alkalinity (mg/l as CaCO3)	Manganese (µg/l)	Iron (µg/l)	Colour (Hazen)
Dunmail raise (TA)	7.2	21	13	33.6	57	9.2
Haweswater (HA)`	7.2	23	14	44.0	81	13.1
Haweswater (HA) Windermere	7.4	21	22	33.6	57	28.0
Haweswater (HA) Wet sleddale	6.9	19	10	111	304	44.0
Haweswater (HA) shap	7.4	30	22	72	75	20.0
Vyrnwy (Oswestry inlet)	6.9	11	6	43	150	25.5
Sutton Hall Intake	7.3	13	8	54	142	18.4
Huntington intake	7.1	12	7	34	153	16.8

4.1.3 Vyrnwy system: The Vyrnwy dam in North Wales was built to enhance the capacity of the existing Lake Vyrnwy at the latter end of the last century, in order to supply the Liverpool metropolitan area. The aqueduct which carries water to the terminal storage reservoirs at Prescott is 109 km long. It conveys 205 Ml/d.

Raw water is conveyed to Oswestry, in Shropshire, where it undergoes treatment. There are twenty three slow sand filters used. Seasonal colours in the raw waters are removed by means of ozonation. After filtration the water is disinfected with chlorine and is pH adjusted with lime dosing. En route to Prescott there are several takeoffs.

4.2 The River Dee

The River Dee provides most of the surface raw water used in NWW region. This is a regulated river which NWW uses along with other abstractors e.g., Hyder, Chester Waterworks Company, British Waterways Board and Dee Valley Water Company. The two major NWW treatment works which use the River Dee are Huntington (400Ml/d) and Sutton Hall (125Ml/d). Neither works has bankside storage, and so vulnerable to raw water quality variation through both season and weather.

Both works employ coagulation, sedimentation and clarification using a mix of aluminium and ferric sulphate called ferral, pH correction with lime and addition of polyelectrolyte. Post clarification, the water is subjected to rapid gravity filtration (RGF) using granular activated carbon (GAC) retero-fitted in the RGF shells to improve the taste of the water and remove pesticides. After filtration the water undergoes disinfection with chlorine and pH adjustment with lime.

Sutton Hall supplies most of the demand for water on the Wirral. Huntington supplies the Dee Aqueduct, via twin pipelines of 60" & 72". These terminate at Prescott reservoirs. En route the Dee Aqueduct supplies Runcorn, Widnes, Warrington, Speke and St. Helens.

4.3 Borehole sources

The boreholes used in the region are mainly in Cheshire and East Lancashire. They are not thought to be a major risk in relation to THM and DBP formation.

5 OPTIONS AVAILABLE TO MINIMISE THE RISK OF BREACHING THE NEW DIRECTIVE THM STANDARD

Action to reduce THM formation has to be balanced with other water quality issues at any given site. Effective disinfection, precipitation of dissolved metals during treatment, water stability and plumbosolvency of the final waters must be given due regard. Other water quality problems faced by the major sources so far discussed will need to be borne in mind when weighing up the best use of financial resources. These include., protozoan parasites, pesticides, algae, tastes and odours. Planning restrictions are also a consideration as many of the assets are located in protected land area e.g., National parks.

The assessment of options below is generic for all sites discussed, however the different operational nature of the upland and surface water assets is reviewed separately.

Optimisation of existing disinfection systems has been undertaken but cannot secure compliance given the nature of the raw waters and the current level of treatment.

5.1 The options available for THM reduction

These fall broadly into three categories:
    removal or destruction of the organic precursors of THMs
    alternative disinfectants which do not form THMs
    removal or destruction of THMs once they have formed
There are a wide range of possible solutions available (Tables 2, 3 and 4).

Table 2 *Removal or destruction of the organic precursors of THMs*

	Process option	Capex	Opex	Other water quality benefits
1.	Ozonation followed by adsorption onto GAC	V.high	High-V.high	Reduce colours, tastes and odours
2.	Ozonation followed by slow sand filtration with GAC sandwich	V.high	V.high	Reduce ,colours, protozoan parasites, bacteria and turbidity
3.	Ozonation followed by coagulation and filtration	High	V high	Reduce colours, protozoan parasites, turbidity and disinfection load
4.	Adsorption of colour onto activated carbon	High	V.high	Reduce tastes and colours
5.	Adsorption of colour onto synthetic resins	High	V.high	Reduce colours
6.	Biosorption of THM precursors	V.high	High	Unknown
7.	Coagulation followed by DAF and filtration	High	Medium	Reduce colours, protozoan parasites, turbidity and disinfection load
8.	Microstraining followed by coagulation and filtration	High	Medium	Reduce colours, protozoan parasites, turbidity and disinfection load
9.	Coagulation and clarification and filtration	High	High	Reduce, colour, protozoan parasites, turbidity and disinfection load

Table 3 *Alternative disinfectants which do not form THMs*

	Process option	Capex	Opex	Other water quality benefits
1.	Ozone	High – V.High	High	Reduced colour(but needs a downstream biological stage to be effective and not create other problems e.g., biofilm)
2.	Chloramines	Low	Low	None
3.	Chlorine dioxide	Low	Low	None

Table 4 *Removal or destruction of THMs once they have formed*

	Process option	Capex	Opex	Other water quality benefits
1.	Adsorption onto activated carbon	High	V.High	Reduction in tastes
2.	Air stripping	High	V.High	None
3.	Biosorption using a bioreactor	V.High	V.high	Unknown

6 BROMATE

30% of our sites use hypochlorite and are expected to struggle to meet the proposed standard The site where Ozonation is used and the sites where on site electrolytic chlorination is practised are not thought to pose a problem.

6.1 Options available to minimise the risk of breaching the new Directive Bromate standard

If commercially available hypochlorite cannot be purchased without bromate as an impurity, the favoured option is to employ onsite electrochlorination with low bromide salt at most sites using hypochlorite . In most cases it a safer option than chlorine gas providing the hydrogen produced can be disposed of safely.

7 APPLICABILITY OF OPTIONS TO SECURE THM COMPLIANCE:

The options are reviewed in Table 5.

7.1 Upland Sources

These sources and the aqueducts which convey them are an impressive and valuable asset resource for NWW. They have been in service for over 150 years and provide a substantial portion of our total resources. Their pivotal role in NWW is a lasting tribute to the foresight of the Liverpool and Manchester corporations who conceived and built them. Their treatment works are however in need of modernisation to meet the new Directive THM standards.

7.1.1 TA system: As the TA system is prone to ground water ingress and has problems with raw water dissolved metals, the aqueduct should therefore be viewed as a raw water source and so treating the water at source is not the most efficient location to reduce THM formation, precipitation of dissolved metals and protozoan parasite risks.

It has numerous takeoffs of varying size and the most effective treatment would be to rationalise these, treat them as raw sources and install appropriate treatment there.

The TA system supplies water to areas where free chlorine is the residual disinfectant used. Hence there is a very real risk of generating dichloramine and nitrogen trichloride with their attendant taste and odour problems if such a strategy was adopted in isolation This is also the case for the Vyrnwy and Haweswater systems and other surface water sources that have been studied. Adopting such a strategy would need to be assessed hollisticaly as the NWW system is very highly integrated.

7.1.2 HA system: The existence of RGF filters and the aqueduct integrity makes the choices more straightforward here

7.1.3 Vyrnwy system: The presence of the existing slow sand filters and ozone plant favours option 2 ,Table 2 here.

7.2 Surface water sources

Both Sutton Hall and Huntington already have coagulation and sedimentation plants. At both sites the existing GRIEFS have been filled with GAC in place of sand. The most cost effective solution would be to go for option 1 in Table 4 and retero-fit the existing

Table 5 *Applicability of options to secure THM compliance*

Source	Option	Reasons
TA	Coagulation and clarification followed by filtration	Expensive but addresses other factors too such as protozoan pathogen risks, colour, turbidity and disinfectant loading and removal of dissolved metals
	Coagulation followed by membrane filtration	More expensive but addresses same factors as above option
	Chloramines	Cheap, possible taste problems Would need to be part of a regional strategy
	Chlorine dioxide	As for chloramines
HA	Microstraining followed by Coagulation and filtration	Expensive but addresses other factors too such as protozoan pathogen risks, colour, turbidity and disinfectant loading and removal of dissolved metals
	Chloramines	As for TA
	Chlorine dioxide	As for TA
Vyrnwy	Ozonation followed by GAC sandwich	Expensive but addresses other factors too such as protozoan pathogen risks, colour, turbidity and disinfectant loading s
	Chloramines	As for TA
	Chlorine dioxide	As for TA
Sutton Hall and Huntington	Adsoption onto GAC	Expensive but addresses other factors too such as dissolved metals, pesticides and tastes
	Chloramines	As for TA
	Chlorine dioxide	As for TA

GRIEFS with sand. These sand filters could be used for manganese removal after suitable pH correction upstream.

8 CONCLUSIONS

The nature of source waters and treatment means that without suitable investment non compliance with the new Directive is certain. Treatment solution for THMs must be considered in balance with other problems, e.g., lead, protozoan pathogen risk, disinfection, tastes, odours colours and dissolved metals. Options vary from source to source dependent on assets already available, land and planning constraints. Significant capital and operation costs are inevitable.

REFERENCES

1. United Utilities Book of Maps
2. PR99 Information Requirement D Main cost of Quality Submissions (1998) OFWAT
3. Water Supply (Water quality) Regulations, S.I 1147. HMSO.

Balancing Chemical and Microbiological Risk

THE RISKS OF DBPs IN PERSPECTIVE

J. K. Fawell

WRc National Centre for Environmental Toxicology
Henley Road, Medmenham,
Marlow SL7 2HD

1 INTRODUCTION

In the 1970s the presence of chloroform was detected in drinking water in Holland[1]. Subsequently, other trihalomethanes (THMs) were also identified. The same period saw the demonstration that chloroform caused an increase in tumours in long-term studies in laboratory animals. This combination of factors was the stimulus for a number of epidemiological studies looking for an association between chlorinated drinking water and various cancers. The importance of chlorination as a water treatment process and its efficacy against a range of important microbiological agents of waterborne human disease required that these data be very carefully considered before any action was taken. Subsequently other disinfectants such as ozone and chlorine dioxide were also shown to give rise to by-products and these too required study. The science of risk assessment in toxicology was developing at the same time and the difficulties encountered in estimating the risks to health from these by-products reflected the development of knowledge in this field. Initially chemicals were thought to be one of the major factors in human cancers and that there was no threshold to chemical carcinogenicity. As knowledge of the biological processes associated with toxicity has increased, so the assessment of risk has also been refined and a more realistic approach has developed.

2 CHLORINATION

2.1 Cancer Epidemiology

A number of epidemiological studies of populations exposed to chlorinated drinking water for long periods of time have been carried out since the mid 1970s, mostly, but not exclusively, in North America. These studies have shown an association between drinking chlorinated water and a variety of cancers, the most frequent being colon, rectum and bladder[2,3,4]. The majority of these studies are of the descriptive, or ecological type, which analyse data which characterise groups or populations rather than individuals. These studies require careful interpretation because of variations in study quality. In particular, it is difficult to fully account for other risk factors for the cancers in question. There have also been a number of case control studies in which individuals from the same area, with or without a specific cancer, were compared using data from death certificates. These studies showed some association with increased colon, rectal or bladder cancer, but the increased risks were very small and each study suffered from limitations[4.]

Several case-control studies, in which detailed information was obtained on each individual, have also been carried out[4,5]. Positive associations were found in various studies with colon, rectal and bladder cancer but these remain weak to moderate with most relative risks below 2.0 and the mean below 1.5, which is only a weak association. The data on bladder cancer are consistent, but the findings on colon and rectal cancer lack consistency. One case control study found a positive association with brain cancer but there are no other studies of this site[5]. All of the studies have some limitations in design or methodology but the two biggest problems are comparing chlorinated surface waters with unchlorinated groundwater where there are other differences than DBPs, and measuring exposure. Concentrations of DBPs have changed over time, with concentrations largely having reduced and the best correlations are associated with long exposures, following the introduction of standards for THMs and better treatment technology. Information on concentrations of DBPs, apart from THMs, are extremely limited even today.

In order to establish causality from epidemiological studies, a number of criteria must be met. These include strength of association, consistency, biological plausibility and dose response relationship. The epidemiological studies of chlorination and cancer do not provide sufficient strength to infer a causal relationship between drinking chlorinated water and cancer. The strength of association is only weak to moderate, consistency between studies is high for bladder, but only moderate for rectum and colon, biological plausibility is only limited as discussed below, and the dose response relationship is weak. In a meta-analysis of all the available studies, Morris[3], showed that overall there were positive associations with bladder and rectum, but not colon. The relative risks for bladder were 1.21 (1.09 - 1.34) and for rectum 1.38 (1.01 - 1.87). In a more recent case control study in Canada there was an association with bladder and colon but not rectum[6]. The epidemiological data do not, therefore, indicate that the risk of cancer associated with drinking chlorinated water is high.

2.2 Toxicological Data

A number of studies have been carried out in laboratory animals to study the carcinogenicity of DBPs from chlorination. The most intensively studied is chloroform but chloroform, bromodichloromethane (BDCM), bromoform, trichloroacetic acid (TCA), dichloroacetic acid (DCA), chloral and MX have all been shown to increase the number of tumours at various sites, in laboratory rats or mice in lifetime experiments[7,8]. The most common sites are liver and kidney, BDCM and bromoform have been shown to increase tumours of the large intestine in rats but none show any activity in bladder. The correspondence with the epidemiological sites is, therefore, limited. In addition, studies of the THMs carried out using corn oil as a dosing vehicle show significantly greater toxicity and carcinogenic potency than those carried out using drinking water as a vehicle. The studies of brominated THMs which gave rise to increases in tumours of the large intestine, were carried out using corn oil but a new, as yet unpublished, study on BDCM using drinking water does not indicate any such increases[9].

In assessing the risks of chemicals, a comparison must be made between the doses which give rise to effects in laboratory animals and the levels to which consumers are exposed. For most toxic endpoints, a margin of safety between 100 and 1000 is normally considered 'safe' and in some cases smaller margins of safety can be tolerated. However, for carcinogenicity this may not be the case. It was considered that substances which caused cancer by interacting with the DNA of a cell to induce mutations, that could be passed on to daughter cells when the cell divided, theoretically had no threshold below which there would be no effect. It was also considered at one time that all chemical carcinogens operated in this way. As a consequence theoretical mathematical models were developed to estimate the risks of cancer at low doses from the risks in animal studies using very high concentrations[10]. This procedure has been used by a number of authorities including the USEPA and WHO. However, although there are differences in the models, the level of risk normally considered acceptable is 1 additional cancer per 100,000 population over a period of 70 years. The value taken from the model is also usually the 95% upper bound and so incorporates a high degree of conservatism, in addition to the conservatism built into the assumptions used in the model. The EU use a risk value of 10^{-6} or 1 additional cancer per million population, giving rise to significantly greater conservatism. However, it is now recognised that not all substances which cause cancer in laboratory animals do so by a direct mechanism and that other, indirect, mechanisms may have a threshold which would render the use of the mathematical models such as the linearised multistage model, inappropriate.

A considerable amount of research effort has been directed at determining the mechanism by which substances such as the THMs and the chloroacetates cause an increase in tumours, There is now substantial evidence that chloroform only increases tumours at doses which are high enough to cause cell death and increased cell division as a repair process[11]. This mechanism will clearly have an experimentally demonstrable threshold which has been acknowledged by both WHO and USEPA. WHO have determined a guideline value for chloroform of 200 μgl^{-1} by applying an uncertainty factor of 1000 to the no observed adverse effect level in a dog study[12]. The data on the brominated THMs are less extensive, but they do not easily damage DNA in studies using whole animals and there are significant uncertainties as to whether they really do cause an increase tumours when given in drinking water. The application of the linearised multistage model to determine the risk of cancer from these substances would also seem inappropriate. The WHO guideline for BDCM is 60 μgl^{-1}, derived using the linearised multistage model, but the use of a threshold approach would give rise to a health based value greater than this.

Di and trichloroacetic acid and chloral have only been shown to increase liver tumours and at extremely high doses. None appear to easily damage DNA in whole animals and all three appear to operate by a mechanism which has a threshold. This is reflected in the WHO guidelines of 50 μgl^{-1}, 100 μgl^{-1} and 10 μgl^{-1} respectively. However the guidelines were all designated as provisional because of limitations to the toxicological database. Subsequently ILSI[11] evaluated dichloroacetic acid on behalf of the USEPA and concluded that " The margin between daily human exposure estimates from drinking water and doses which induce toxicity (and neoplasia) in experimental animals is large (approximately 1300). This ratio is believed to be sufficient to encompass factors relating to variability." The panel also went on to indicate that they "did not believe that there was cause for immediate concerns regarding public health risks of dichloroacetic acid exposures to human populations at current levels in drinking water."

Other by-products of chlorination are present at much lower concentrations and, on the whole, are less well studied. The substances of greatest note are the haloacetonitriles and the chlorinated furanone, 3-chloro-4-(dichloromethyl)-5-hydroxy-2(5H)-furanone (MX). The haloacetonitriles have been shown to variously possess mutagenic properties *in vitro* and to cause malformations of the foetus in rats[12.]. However, the concentrations encountered in drinking water are low and there appears to be a substantial margin of safety. MX is a potent mutagen in bacterial tests *in vitro* and accounts for the great majority of the mutagenicity in bacterial tests found with concentrated extracts of chlorinated drinking water[13]. It has apparently been shown to increase tumours at a number of sites in rats given very high doses[8]. None of these sites coincides with the sites identified in the epidemiological studies and the data on metabolism and tissue binding, using radioabelled MX do not fit well with the data from the carcinogenicity study[14]. The ILSI workshop on the health effects of chlorination by-products held in North Carolina in 1996[15] concluded that the levels of exposure to MX of less than 100 ngl^{-1} were so low that this was not a high priority for further research. An editorial in the issue of the Journal of the National Cancer Institute in which the results of the MX carcinogenicity study were published also concluded that the risks of cancer from exposure to MX in drinking water were well within acceptable levels[16].

2.3 Reproductive Effects

The most recent studies relating to exposure to the by-products of chlorination are those which have examined effects on reproductive outcome in human populations. Since the critical period for measurement of exposure is the period of gestation, and is, therefore, relatively short, it should be possible to characterise exposure much more accurately than in cancer studies. However, only one study to date has used very detailed measurement of individual exposure. Associations with neural tube defects, cardiac anomalies, oral cleft defects, still birth, miscarriage, preterm delivery, growth retardation and low birth weight have been reported[17,18]. However, there are only a small number of studies, the consistency is limited and the relative risks are weak to moderate. The most convincing associations, so far, are with neural tube and central nervous system defects, and with low birthweight, stillbirth and miscarriage. However, the data are insufficient to imply causality and there is a need to carry out more extensive and very well conducted studies in a range of locations. There is also a need to examine the possible mechanism for reproductive defects and the plausibility of such effects occurring as a consequence of exposure to the low concentrations of DBPs found in drinking water. The implication is that the agent, or agents, are extremely potent.

At present there is no evidence to support this hypothesis. The data on reproductive effects are insufficient to consider stopping chlorination, but they are sufficient to require further more detailed investigation and must be taken seriously.

3 OZONATION

3.1 Health Effects

The use of ozone is relatively recent and it is most commonly, though not exclusively, used in conjunction with a second disinfectant, such as chlorine, which will provide a residual level of activity in distribution. It would appear that there are no epidemiological studies available which have examined health effects in populations exposed to exclusively ozonated water. The by-products of ozonation which have excited most interest, so far, are formaldehyde and bromate which were examined by WHO in their 1993 Guidelines for Drinking-Water Quality[19]. The concern has been carcinogenicity, formaldehyde because of the evidence of nasal carcinogenicity following inhalation and bromate because of data in laboratory animals.

WHO considered formaldehyde in their revision of the Guidelines for Drinking-Water quality[7,19] and concluded that there was no evidence of carcinogenicity by the oral route. They recommended a guideline value of 900 μgl^{-1}. Subsequently laboratory animal studies on formaldehyde and acetaldehyde have claimed to demonstrate the induction of tumours at several sites in laboratory rats[20] but details of the studies are limited and there are a number of inconsistencies which cast considerable doubt on their significance. Mechanistic data on the toxic action of formaldehyde and acetaldehyde also support the position that there is no risk associated with the low concentrations found in drinking water.

Bromate has been shown to cause an increase in tumours of the kidney, peritoneal mesotheliomas and follicular cell tumours of the thyroid[9,21] in rats and to a much lesser extent in hamsters, but not mice. Bromate also causes chromosome damage *in vitro* and *in vivo* but does not cause mutations in bacteria[14]. The chemical characteristics of bromate do not suggest a direct genotoxic mechanism of action and although WHO calculated an excess cancer risk using the linearised multistage model, they indicated that if the mechanism was by oxidation the use of such models would be inappropriate[19]. Subsequently, a significant body of evidence has been collected to show that the kidney tumours are secondary to oxidative damage to cell membrane lipids and that this is a high dose phenomenon[22,23]. This indicates that the risk associated with the low concentrations of bromate found in drinking water are unlikely to pose any risk to human health.

4 CHLORINE DIOXIDE

4.1 Health Effects in Man

Chlorine dioxide has been used as a drinking water disinfectant since the 1940s and the concerns have centred round the inorganic breakdown products, chlorite, which is the major one, and chlorate. There have been both epidemiological studies and clinical trials in which known concentrations of chlorine dioxide, chlorite and chlorate were given to human volunteers. The primary concern about chlorine dioxide and its breakdown products in these studies was effects on the red cells.

Clinical studies were carried out in healthy volunteers for periods of up to four weeks in which blood samples were taken and analysed for a range of parameters. These studies included a number of individuals with the inherited genetic deficiency of the enzyme glucose-6-phosphate dehydrogenase, who are more susceptible to oxidative damage to the blood cells. However, no adverse effects were found at the highest dose 0.036 $mgkg^{-1}$ body weight.

A limited number of epidemiological studies have been carried out on populations exposed to chlorine dioxide treated drinking water[24,25,26]. These have examined blood parameters and measures of reduced size at birth. All of these studies suffer limitations. There were some indications of reduced size at birth, and an increase in neonatal jaundice in one study but the limitations of the studies do not allow any firm conclusions to be drawn, particularly in view of the fact that most of the differences were small.

4.2 Toxicity Studies

Early studies on laboratory animals with chlorine dioxide and its breakdown products indicated possible effects on red blood cells, the thyroid and neurodevelopmental effects in rats[7]. Subsequently, a number of other studies have provided no evidence of significant haematological effects in the rat, mouse or monkey[7,27] at concentrations which would give rise to concern for consumers. These include modern toxicological studies carried out to the highest standards and also provide no supportive evidence for effects on the thyroid or on neurodevelopment.

The evidence from mutagenicity studies is that neither chlorite or chlorate are mutagenic and chlorite showed no carcinogenic potential in long-term studies in laboratory animals[7]. A long-term carcinogenicity study on chlorate is planned in the USA, but there was no evidence of effects in a tumour promotion study on rat kidney[21].

As a consequence of the new data generated in support of the risk assessment of chlorine dioxide, it would appear that chlorine dioxide under current conditions, is unlikely to give rise to any significant risk to public health. Indeed the USEPA are proposing to increase the health based maximum contaminant level goal to 0.8 mgl[-1] for both chlorine dioxide and chlorite. Chlorate has not been specifically mentioned in the new proposals because chlorate concentrations are much lower than chlorite concentrations, and have stated that they consider their maximum contaminant level of 1.0 mgl[-1] provides an adequate margin of safety for the whole population[28].

5 DISCUSSION AND CONCLUSIONS

Chlorination is the best studied of the major disinfection processes. Although associations with cancer and reproductive effects have been identified in epidemiological studies, it remains uncertain whether any relationship with drinking water chlorination is causal. However, the evidence on reproductive effects is sufficient to merit further research, as long as this research is of an appropriate high quality. The data on ozone are much more limited but only one major issue, the carcinogenicity of bromate, has so far emerged. New data on the mechanism of carcinogenicity in laboratory animals indicates that this unlikely to be of concern at the concentrations encountered in drinking water. There are limited epidemiological data on chlorine dioxide and its breakdown products but the toxicological database is, mostly, very good. These new data have helped to remove much of the uncertainty surrounding chlorine dioxide and provide a basis for confidence in its safety as a drinking water disinfectant.

The evidence, therefore, supports the view that although there are data on possible health effects associated with chlorination, causality is uncertain and the risks appear to be small, especially in view of steps taken to improve chlorination practice to reduce the levels of exposure to chlorination by-products. Although the other disinfectants are less well studied, there are no indications, so far, of any significant risks to health. By contrast, the risks to health of microbiological contaminants are significant and there appears no reason why the use of the disinfectants discussed here should not remain an important component of the multibarrier approach to water treatment which has served well over many years.

References

1. J. J. Rook, *Wat. Treatment Exam.*, 1974, **23,** 234.
2. International Agency for Research on Cancer, Chlorinated drinking-water, chlorination by-products, some halogenated compounds, cobalt and cobalt compounds. IARC Monographs on the evaluation of carcinogenic risks to humans No 52. IARC, Lyon 1992.
3. R. D. Morris, A. M. Audet, I. F. Angelillo, T. C. Chalmers and F. Mosteller. *Ann. J. Publ. Hlth.* 1992, **82,** 955.
4. G. F. Craun, 'Safety of Water Disinfection: Balancing Chemical and Micrbial Risks' Ed. G. F. Craun. ILSI Press, Washington D. C. , 1993, p. 277.
5. K. P. Cantor, *Cancer Causes and Control.* 1997, **8,** 292.
6. L. D. Marrott and W. D. King, 'Great Lakes basin cancer risk assessment; a cse-control study of cancers of the bladder, colon and rectum. Research Report prepared for the Bureau of Chronic Disease Epidemiology Laboratory, Centre for Disease Control, Health Canada. 1995,
7. World Health Organization. Guidelines for drinking-water quality. Second edition. Volume 2. Health criteria and other supporting information. World Health Organization, Geneva, 1996.
8. H. Komulainen, V-M. Kosma, S-L. Vaittinen, T Vartiainen, E. Kaliste-Korhonen, S. Lotjonen, R. K. Tuominen and J. Tuomisto. *JNCI,* 1997, **89,** p. 848.
9. A. B. DeAngelo. USEPA National Health and Environmental Effects Laboratory. Personal Communication.
10. J. K. Fawell and W. Young. *Ann Ist Super Sanita,* 1993, **29,** p. 313.
11. International Life Sciences Institute. An evaluation of EPA's proposed guidelines for carcinogen risk assessment using chloroform and dichloroacetate as case studies. ILSI, Washington D.C., 1997
12. World Health Organization. Guidelines for drinking-water quality. Second edition. Addendum to volume 2. Health Criteria and other supporting information. World Health Organization, Geneva, 1998.
13. J. K. Fawell and H. H. Horth. Genetic Toxicology of Complex Mixtures. Eds M. D. Waters, F. B. Daniel, J. Lewtas, M. M. Moore and S. Nesnow. Environmental Health Science Research Vol 39. Plenum Press, New York and London, 1990, p. 197.
14. J. K. Fawell, G. O'Neill., J.P. Dunsire and A. M. Johnston, Disinfection by-products in drinking water: Critical issues in health effects research. International Life Sciences Institute, Washington D.C. 1996, p.131
15. International Life Sciences Institute. Disinfection by-products in drinking water: Critical issues in health effects research. Workshop report Chapel Hill, North Carolina,October 23-25, 1995. ILSI, Washington D.C., 1996.
16. R. L. Melnick, G. A. Boorman and V. Dellarco. Editorial: Water chlorination, MX and potential cancer risk. *JNCI,* 1997, **89,** p. 1.
17. J. S. Reif, M. C. Hatch, M. Bracken, L. B. Holmes, B. A. Schwetz and P. C. Singer. *EHP,* 1996, **104,** p. 1056.
18. K. Waller, S. H. Swann, G. DeLorenze and B. Hopkins. *Epidemiology,* 1998, **9,** p. 134.
19. World Health Organization. Guidelines for drinking-water quality. Second Edition. Volume 1. Recommendations. WHO. Geneva, 1993.
20. M. Soffritti. Disinfection by-products in drinking water: Critical issues in health effects research. International Life Sciences Institute, Washington D.C.. 1996, p. 51

21. Y. Kurokawa, A. Maekawa, M. Takahashi and Y. Hayashi. *EHP,* 1990, **87,** p. 309.

22. J. K. Chipman, J. E. Davies, J. L. Parsons, G. O'Neill and J. K. Fawell. *Toxicology,* 1998, **126,** p. 93.

23. J. K. Chipman. This volume.

24. G. E. Michael, R. K. Miday, J. P. Bercz, R. G. Miller, D. G. Greathouse, D. F. Kraemer and J. B. Lucas. *Arch. Environ. Hlth.* 1981, **36,** p. 20.

25. R. W. Tuthill, R. A. Guisti, G. S. Moore and E.J. Calabrese. *EHP,* 1982, **46,** p. 39.

26. S. Kanitz, Y. Franco, V. Patrone, M. Caltabellote, E. Raffo, C. Riggi, D. Timitilli and G. Ravera. *EHP,* 1996, **104,** p.516.

27. Federal Register National primary drinking water regulations: disinfectants and disinfection by-products. Notice of data availability Tuesday March 31 1998.

BROMATE CARCINOGENICITY: A NON-LINEAR DOSE RESPONSE MECHANISM?

J.K. Chipman

School of Biochemistry
The University of Birmingham
Edgbaston
Birmingham
B15 2TT UK

1 INTRODUCTION

A major dilemma with respect to treatment of drinking water is the need to determine a satisfactory balance between the undoubted health benefits of disinfection and the potential risk from harmful disinfection by-products (DBP). This is emphasised by contrasting health concerns recently reported. On the one hand, inadequate disinfection because of the concern of DBP led to a major outbreak of cholera in Peru[1]. Conversely, several epidemiological studies have suggested that a correlation exists between certain human cancers (bladder and rectal) and the practice of water chlorination[2,3,4]. Only with an adequate risk assessment of DBP, can an acceptable balance between adverse and beneficial effects be determined. Epidemiology plays a part of this risk assessment but is often difficult to interpret e.g. due to confounding factors, inadequate controls and insufficient sample numbers. The risk assessment therefore needs to rely very much on studies of adverse effects in experimental animals and other biological systems. It has become increasingly clear that when considering the risks associated with carcinogens, the mechanism of action of such chemicals should be taken into account as well as the concentrations to which individuals may be exposed.

2 THE MECHANISM OF CARCINOGEN ACTION

Cancer is a multi-stage process involving sequential mutational changes in critical genes (proto-oncogenes and tumour suppressor genes) leading to loss of cell growth control[5]. Many carcinogens can contribute to this pathway through their ability to damage DNA (genotoxicity). Following subsequent cell division or by faulty repair, DNA damage can lead to mutations and thus to cancer. Conversely, it is now recognised that chemicals also exist which cause cancer (often with species and target organ specificity) by a non-genotoxic mechanism[6]. Understanding the distinction between these two classes of carcinogen (genotoxic versus non-genotoxic) may be crucial in risk assessment since the dose response relationships may differ dramatically.

For many non-genotoxic carcinogens a threshold-dose appears to be necessary for a carcinogenic effect[6]. Furthermore some non-genotoxic carcinogens such as the peroxisome proliferators exera rodent-specific effect which is considered not to be relevant to human liver. Particular concern is therefore given to genotoxic carcinogens since potentially relatively very low exposures may impose a risk of a critical DNA damaging event to occur and therefore a cancer risk to exist. This dose response relationship will be discussed more fully below in the context of potassium bromate.

3 CARCINOGENIC DBP

A number of DBP from the chlorination of water possess carcinogenic activity in animals. In some cases, however, a non-genotoxic mechanism (see above) appears to exist. Chloroform, for example has been shown to cause liver and kidney tumours in rodents [7], despite an apparent lack of genetic toxicity.

In contrast to chloroform, concern has been given to the level of the furanone (MX). This DBP is clearly genotoxic and mutagenic in a range of bacterial and mammalian cell assays[8]. Although the high reactivity of MX might minimise the absorption of this DBP from the intestine[9,10] into the body, it has very recently been demonstrated that this compound is a multi-organ carcinogen in the rat[11]. This finding (coupled with the known genotoxicity) needs to be considered in the light of the epidemiology on the cancer risk of chlorinated drinking water (see above).

Ozonation of water (though avoiding the formation of trihalomethanes (THMs) and the MX product, is however also not devoid of the formation of animal carcinogens. These include formaldehyde, hydrogen peroxide and bromate. Of these products, particular concern has been given to potassium bromate. Potassium bromate has been demonstrated to cause renal cell tumours, mesotheliomas and thyroid tumours in rats [12]. Supplementary studies have shown the ability of potassium bromate to produce chromosomal aberrations and micronuclei in mammalian cells and mutations in bacteria[13-15]. These data lead to the conclusion that bromate is a genotoxic carcinogen (see above) and is therefore a possible human carcinogen.

4 RISK ASSESSMENT OF BROMATE DEPENDS ON ITS MECHANISM OF ACTION

For the risk assessment of bromate in drinking water it is necessary to predict a "safe" level of intake, however this has to be determined based on results from relatively high dose exposures in the animal carcinogenicity studies. If a "single hit" of a genotoxic carcinogen at a critical site in DNA can contribute to cancer, then it might be argued that there is no "safe" dose of bromate or any other genotoxic carcinogen. However, it is clear that DNA is a target for substantial background (or endogenous) damage including that produced by reaction oxygen[16]. Hence a certain degree of DNA damage from xenobiotic sources may be trivial in this context. Moreover, organisms have developed efficient protective and repair processes which may become overwhelmed at relatively high doses of carcinogens. Consequently, exposures which are considered to be "virtually safe" or an "acceptable risk" need to be established. In the case of potassium bromate, the provisional guideline value advised by the World Health Organisation (WHO) is 25 µg/l in drinking water. The proposed EU directive which is likely to be adopted gives a value of 10 µg/l for bromate [17]. Also in the USA, the EPA are proposing to regulate bromate 10 µg/l [18]. These decisions are based on the practical detection limits for bromate. But what is the risk associated with such exposure? It has been calculated that for a risk of 10^{-5} (1 cancer in 100,000 for a lifetime exposure via drinking water) a level of 3 µg/l (WHO) or 0.5 µg/l (EPA) is appropriate. The difference depends on the model used for extrapolation to low dose[17,18]. To what extent are the models used appropriate for the risk assessment of bromate? Without a detailed understanding of the mechanism of action of bromate, this question cannot be fully answered, however it is particularly useful to consider the dose response relationship in the action of bromate since this may influence decisions on the appropriate model to be used for low dose extrapolation and risk assessment.

5 DOSE-RESPONSE RELATIONSHIPS IN THE ACTION OF BROMATE

A crucial question in attempting to understand the dose-response relationship in bromate carcinogenicity, relates to the mechanisms whereby DNA is damaged. In particular, does bromate damage DNA directly or is DNA damage produced or enhanced following the depletion of cellular antioxidants or reductants by bromate exposure and by the consequent lipid peroxide formation (i.e. a potential "secondary" mechanism). Using isolated DNA and cellular systems, there is good evidence that potassium bromate can cause DNA oxidation directly. This appears to be mediated by a radical species (not identified) produced by the interaction of bromate with the cellular reductant glutathione (GSH)[19,20]. But does this direct mechanism for DNA damage occur *in vivo?* Several points of information suggest that DNA damage by bromate in kidney cells *in vivo* may involve the "secondary" mechanism discussed above. Firstly, extracellular glutathione can protect against DNA damage by bromate both *in vitro* and *in vivo*[20,21,22]. Interestingly, in relation to this, bromate is excreted from the body more as the reduced form (bromide) at relatively low doses than at higher doses when bromate itself appears in the urine[12]. The protection by GSH may reflect formation of the damaging radical species outside of the cell, limiting access to DNA. Furthermore, Sai *et al.*[21] found that optimal DNA oxidation (8-hydroxy deoxyguanosine (8-0HdG) occurred with a delay of 24 hours after bromate administration to rats by intraperitoneal injection. This DNA oxidation was seen at a dose of bromate which was also associated with lipid peroxidation in the kidney and evidence of kidney toxicity (increase in relative kidney weight)[21]. We have also found the level of both 8-OHdG and etheno-DNA adducts (from lipid peroxides) not to be elevated in rat kidney at a dose of bromate which caused no lipid peroxidation or GSH oxidation[23]. 8-OHdG was, however, elevated in association with the markers of oxidated stress (GSSG and lipid peroxide) at 100 mg/kg (i.p.). Although these results do not necessarily indicate a threshold-dose for DNA damage in rat kidney, they suggest the likelihood of a non-linear dose response. Such non-linearity is also suggested from the incidence of renal cell tumours in rats treated with potassium bromate[24]. This study demonstrated a sigmoid dose-response relationship and the authors suggested a virtually safe dose of 0.950 mg/l for renal cell tumours using a probit analysis which gave a good fit. Other models gave substantially lower values for a virtually safe dose but with lower p values. The various studies on the effects of potassium bromate indicate that the mechanism of action and dose-response relationship is complicated and is not well understood. However the apparent non-linearity in response should be considered in risk assessment. In addition to the need for a better understanding of the mechanism of action, it would be beneficial to have information from a well-controlled epidemiological study on the effects of water ozonation on cancer incidence.

ACKNOWLEDGEMENTS

The work in the author's laboratory has been in collaboration with Geraldine O'Neal, Wendy Young and John Fawell of the W.R.c, Medmenham, U.K. and partly supported by UKWIR.

REFERENCES

1.   A.A. Ries, D.J. Vugia, L. Beringolea, A.M. Palacois, E.Vasquez, J.G. Wells, N, Garcia, D.L. Swerdlow, M. Pollak, N.H. Bean, L. Seminario and R.V. Tauxe. *J. Infectious Diseases*, 1992, 166, 1429.
2.   R.D. Morris, A.M. Audet, I.F. Angelillo, T.C. Chalmers and F. Mosteller. *Am. J. Public Health*, 1992, 82, 955.
3.   K.P. Cantor. *Am. J. Public Health*, 1994, 84, 1211.
4.   D.M. Freedman, K.P. Cantor, N.L. Lee, L.S. Chen, H.H. Lei, C.E. Ruhl and S.S. Wang. *Cancer Causes and Control*, 1997, 8, 738.
5.   B. Vogelstein and K.W. Kinzler. *Trends in Genetics*, 1993, 9, 138.
6.   I. Purchase. *Human Exptl. Toxicol,* 1994, 13, 17.
7.   J.K. Dunnick and R.L. Melnick. *J. Natl. Cancer Inst.*, 1993, 85, 817.
8.   J.R. Meier, W.F. Blazak and R.B. Knohl. *Environ. Mol. Mut.*, 1987, 10, 414.
9.   N.W.E. Clark and J.K. Chipman. *Toxicol. Lett.* 1995, 81, 33.
10. J.W. Nunn, J.E. Davies and J.K. Chipman. *Mutation Res*, 1997, 373, 67.
11. H. Komulainen, V.-M. Kosma, S.-L.Vaittinen, T. Vartiainen, E. Kaliste-Korhonen, S. Lötjönen, R.K.Tuominen and J. Tuomisto. *J. Natl. Cancer Inst.*, 1997, 12, 848.
12. Y. Kurokawa, A. Maekawa, M. Takahashi and Y. Hayashi. *Environ. Health Perspect.*, 1990, 87, 309.
13. M. Ishidate Jr., T. Sofuni, K.Yoshikawa, M.Hayashi, T. Nohmi, M. Sawada, and A. Matsuoka. *Fd. Chem. Toxicol.*, 1984, 22, 623.
14. M. Ishidate Jr. and K.Yoshikawa. *Arch. Toxicol. Suppl.*, 1980, 4, 41.
15. M. Hayashi, M. Kishi, T. Sofuni and M. Ishidate Jr. *Fd. Chem. Toxicol.* 1988, 26, 487.
16. B.N. Ames and L.S. Gold. *Mutation Res.*, 1991, 250, 3.
17. O. Hydes, O. DBP standards and regulation. The UK and European situation. Proc. Int. Conf. on Disinfection Byproducts - The way forward. *The Royal Soc. Chemistry*, 1999.

18. S. Regli, M. Cox, T. Grubbs. and J. McLain. DBP standards and regulation: the North American situation. Proc. Int. Conf. on Disinfection Byproducts - the way forward. *The Royal Soc. Chemistry*, 1999.

19. D. Ballmaier and B. Epe, B. *Carcinogenesis*, 1995, 16, 335.

20. J.L. Parsons. and J.K. Chipman. *Human Exptl. Toxicology* Abstract. In Press.

21. K. Sai, A. Takagi, T.Umemura, R. Hasegawa. and Y. Kurokawa. *Jpn. J. Cancer Res.*, 1991, 82, 165.

22. K.Sai, T. Umemura, A.Takagi, R.Hasegawa and Y. Kurogawa. *Jpn. J. Cancer Res.*, 1992, 83, 45.

23. J.K. Chipman, J.E. Davies, J.E. Parsons, J.L. Nair, G. O'Neill and J.K. Fawell. *Toxicology*, 1998, 126, 93.

24. Y. Kurokawa, S. Aoki, Y. Matsushima, N. Takamura, T. Imazawa and Y. Hayashi. *J. Natl. Cancer Inst.*, 1986, 4, 977.

DEVELOPMENTS IN MICROBIOLOGICAL RISK ASSESSMENT: FROM *CRYPTOSPORIDIUM* TO BSE

P. Gale

WRc plc,
Henley Road, Medmenham,
Marlow, Bucks, SL7 2HD, UK.

1 INTRODUCTION

Pathogens can gain entry to drinking water supplies. Break-through during treatment and ingress through cracked pipes are well-documented causes of waterborne outbreaks. Microbiological risk assessment (MRA) is the emerging method to predict the risks to public health from those pathogens. The approach has been used to set microbial standards. It could also be used to answer questions such as how many more people will be infected if part of drinking water treatment fails and by how much will public health be jeopardised if disinfection is eliminated.

There are many difficulties and imponderables in MRA for drinking water. Key questions to address in developing MRA models for *Cryptosporidium parvum* are, "What risks do single oocysts pose to consumers within a given population if ingested?" and, "What doses of oocysts (0, 1, 2, 3, ...10,...100,...) are ingested by individual consumers through drinking water under outbreak and non-outbreak conditions?". It is important to recognise that the final prediction is only as good as the assumptions on which it is based. One problem with quoting quantitative predicted risks is that the degree of uncertainty is quickly forgotten.

1.1 Current MRA models for drinking water- How accurate does risk assessment need to be?

MRA models have been developed for enteric viruses, *Giardia*, and for *C. parvum* in drinking water[1]. Recently a risk assessment approach has been published for BSE in drinking water[2]. The BSE risk assessment is unique because there is no epidemiological evidence to identify its route of transmission from cattle to humans. Indeed it was only recently that it was proved that the 39 cases of new variant CJD in young adults (to 1/1/99) and BSE were indeed caused by the same agent. To restore public confidence in British beef, a huge cull programme of cattle over 30 months old was announced. Over 800,000 cattle were culled in 1996 alone. This raised concerns about BSE agent in the aquatic environment and indeed whether drinking water consumers were at risk. The risk assessment for BSE was based on the barriers which prevent the BSE agent from breaking through into drinking water supplies. The risk of infection estimated for individual consumers drinking $2 \, l \, d^{-1}$ of water from an aquifer potentially contaminated with effluent from a rendering plant was 1.5×10^{-8} person^{-1} year^{-1}. This was based on worst case assumptions and the predicted risk *could* be 10^{20}-fold too pessimistic depending on whether there is a threshold dose for infection by BSE prions[3]. Despite this uncertainty, the risk assessment is still useful because it virtually eliminates drinking water as a route of transmission to humans. However, problems arise when quantitative risks predicted for different sources of exposure are directly compared. For example, quantitative MRAs which do not take into account the very different nature of the exposures to BSE prions through drinking water and beef-on-the-bone actually predict a higher risk of BSE infection through water than through direct consumption of beef-on-the-bone[3]. This illustrates the point that for some MRAs, it is better to be roughly right than precise and wrong.

For known waterborne pathogens, such as *C. parvum*, the risk assessments have to be much more accurate. Indeed to be more than a factor of three-fold too pessimistic would be unacceptable because to improve water quality unnecessarily would be a waste of resource. Risk assessment models are based on an estimate of pathogen exposure. This is calculated as the product of pathogen density in the drinking water and the volume of water drunk by individual consumers. That exposure is then translated into the risk of infection through a dose-response curve.

2 MODELLING PATHOGEN EXPOSURES THROUGH DRINKING WATER

Accurately assessing pathogen exposures through drinking water is not straight-forward. First there is large variation in micro-organism densities in drinking water[1]. Second, most large volume samples record zero pathogens even during outbreaks[4].

2.1 Variation in indicator bacteria concentration in drinking water

Counts of bacteria including total heterotrophic bacteria (22°C and 37°C plate counts) and total coliforms show considerable variation in drinking water[1]. This variation is both spatial and temporal can be modelled to some extent in UK drinking water supplies by changes in chlorine and temperature[5]. In recent years, protozoans and viruses have replaced bacterial pathogens as the primary causative agent of waterborne disease. While conditions of high temperature and low chlorine may promote bacterial growth, protozoa and viruses cannot multiply in the aquatic environment and some, most notably, *Cryptosporidium*, are resistant to chlorine. Therefore, it would be surprising if counts of such pathogens showed the same degree of variation in drinking water as for bacteria.

2.2 Variation in pathogen densities between and within large volume samples

Pathogen density data reported for drinking water supplies under non-outbreak conditions are presented in Table 1. The volumes sampled ranged between 100 and 1,000-l. Most samples contained zero pathogens although some contained high counts. This would be consistent with considerable spatial and temporal variation *between* large volume samples. Most drinking water consumers drink in the region of 1 to 2-l of tap water per day although a considerable proportion of that may be boiled in the UK[1]. This raises the question of how pathogens are distributed *within* those large volumes at the resolution of the volumes ingested daily by drinking water consumers. The highest *Cryptosporidium* count measured in a 100-l volume under non-outbreak conditions was about 50 oocysts (Table 1). Three scenarios could be envisaged with 50 people each drinking 2-l of that 100-l volume:-

1. under-dispersion - everybody gets one oocyst
2. Poisson-dispersion - most people get between 0 and 4 oocysts
3. over-dispersion - in the extreme one person gets all 50 oocysts and the other 49 people get zero.

Table 1   *Pathogen count data in drinking water studies[1]*

Pathogen	Sample Volume (l)	% with zero counts	Maximum count
Enteric viruses	1,000	93	20
Giardia cysts	100	83	64
Cryptosporidium oocysts	100	73	48

Attachment of pathogens to particulates could produce the extreme scenario 3, although there is no evidence for this to date. The available evidence suggests that the effect may be somewhere between scenarios 2 and 3. Thus, Pipes[6] in 1977 analysed 100-ml samples from the same well-stirred 10-l volume of tap water and observed a considerable range of counts. Surprisingly some samples had zero coliforms while others contained over 20 coliforms. The counts did not fit the Poisson distribution which predicted no zeros and no samples with more than about 12 coliforms. The data suggested the presence of 'coliform-poor' and 'coliform-rich' regions within the same volume of water.

2.3 Effect of treatment on the spatial distribution of micro-organisms

The spatial distribution of spores of aerobic spore-forming bacteria in 100-l volumes of water has recently been investigated before and after treatment[7]. The treatment process studied was alum-coagulation followed by rapid gravity sand filtration at an operational drinking water treatment works. The overall conclusion was that while drinking water treatment removed 94 - 98% of aerobic spores it also promoted the spatial association of spores. Thus the spore counts were in general Poisson distributed within raw water volumes (Figure 1). In contrast, the counts of aerobic spores measured in treated water volumes showed much greater variation in magnitude and in general could not be modelled by the Poisson distribution (Figure 2). However, those counts could be better modelled by the negative binomial and Poisson-log-Normal distributions. There are several possible reasons for the spatial heterogeneity observed in micro-organism counts within treated water volumes:-

1. The odd high count samples reflect errors. However, it should be pointed out that high count samples were more often observed in the treated water samples but only rarely in the raw waters.

2. Drinking water treatment may let micro-organisms break-through filters in a non-homogeneous manner.
3. Heterogeneity could reflect micro-organisms released from particles during detergent treatment.
4. Heterogeneity in spore counts could reflect different regrowth rates within the sand filter environment and sporolation of vegetative bacteria.

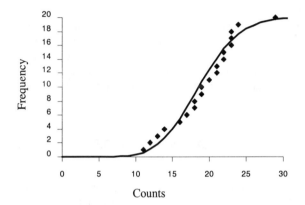

Figure 1 *Cumulative negative binomial (faint line - obscured) and Poisson (bold line) frequency distributions for aerobic spore counts in 0.1-ml samples taken from a 100-l raw water volume.*

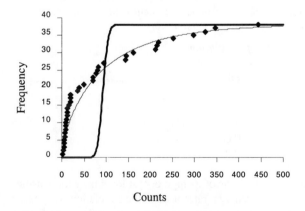

Figure 2 *Cumulative negative binomial (faint line) and Poisson (bold line) frequency distributions for aerobic spore counts in 100-ml samples taken from a 100-l treated water volume.*

In addition there is the question of whether the effect is 'species-specific' (i.e. some micro-organisms show heterogeneity, while others do not) or 'environment-specific' (i.e. the degree of spatial heterogeneity is related to water-type and treatment). Pilot plant experiments (unpublished) suggested that the spatial heterogeneity of vegetative bacteria (total coliforms and 22°C plate counts) was little affected by drinking water treatment. Of particular relevance is the question of how oocysts of *Cryptosporidium* are distributed within treated drinking water samples.

The spatial heterogeneity observed in treated water samples for spores provides a potential mechanism whereby a few consumers could be exposed to doses of more than one oocyst in a single exposure - even during non-outbreak conditions. It could be argued that pathogen densities are too dilute in drinking water for this to be possible, but as shown in Table 1 a few samples do exhibit high counts. Furthermore, other rarer samples with still higher counts could have been missed particularly in view of the log-Normal distribution of pathogen densities in treated water (see below).

3 SIMULATING *CRYPTOSPORIDIUM* EXPOSURES UNDER OUTBREAK CONDITIONS

Doses of *Cryptosporidium* oocysts ingested daily by individual consumers through drinking water were simulated by Monte Carlo methods[8] (Table 2) to take into account the variation in:-

1. pathogen densities in raw water;
2. treatment removal efficiency; and
3. daily tap water consumption by individuals.

Table 2   *Poisson-log-Normal distribution for doses of Cryptosporidium oocysts ingested daily by individual consumers under conditions consistent with a waterborne outbreak. Arithmetic mean exposure is 0.37 oocysts person^{-1} day^{-1}.*

Dose (Oocysts person^{-1} day^{-1})	% of population exposed to dose
0	88.4
1	6.7
>5	1.2
>10	0.6
>50	0.06
>100	0.01

The raw water data set was chosen on the basis of its high oocyst loadings with densities ranging between about 1 and 100/l. Oocyst removal rates used in the simulation were based on particle removal rates reported for drinking water treatment plants and ranged from 1 $\log_{10}$ to 5 $\log_{10}$. The net removal rate in the simulation as represented by the arithmetic mean was 1.96 $\log_{10}$ units and was much less than suggested by the geometric mean removal rate (or average log removal) of 2.7 $\log_{10}$ units. The simulated oocyst densities in the drinking water varied between <1/10,000-l and 164/l. The highest exposure simulated was 98 oocysts person[-1] day[-1]. Using the dose-response curve for *C. parvum* in healthy adult volunteers[9] (Figure 3), the exposures simulated in Table 2 would result in about 15 infections /10,000 persons/day. This would represent an outbreak.

Table 3 *Poisson-log-Normal distribution for counts of Cryptosporidium oocysts in 100-l volumes of treated water simulated for conditions consistent with an outbreak*

Oocyst count	% of samples
0	32.9
1	13.8
>1	53.3
>5	32.6
>10	23.4
>30*	12.0
>100	4.7
>1,000	0.5
>5,000	0.02

*arithmetic mean

Using the same Monte Carlo approach as for Table 2, counts of oocysts in 100-l volume 'spot' samples taken from the supply were simulated (Table 3). Counts showed great variation. While a third of samples contained zero oocysts, a small proportion contained hundreds and even thousands of oocysts. Clearly, a recording of zero oocysts/100-l offers little reassurance and sampling programmes should be designed to catch the rare but all important high count samples. In this respect, continuous sampling offers advantages over taking single 'spot' samples. The simulated counts in Table 3 resemble those reported in recently-documented waterborne outbreaks of cryptosporidiosis in the UK and USA[4]. In some of those outbreak no oocysts were ever detected. While 66% of samples recorded 0 oocysts /1,000-l in the Farmoor (UK) outbreak, one sample was reported to contain 77,000 oocysts/1,000-l. A consumer ingesting 2-l of that water would have ingested about 150 oocysts in a single exposure. The oocyst exposures simulated in Table 2 may be used as a model for a waterborne outbreak. Most people are not exposed to any oocysts each day. Of those who are exposed about half ingest just one a day, while some ingest higher doses. A very small proportion (0.01%) ingest more than 100 oocysts in a single day during the outbreak. The ID_{50} for *C. parvum* is about 130 oocysts (Figure 3).

3.1 Ignoring the variation would predict that more consumers are exposed, albeit to lower doses

The average exposure for the simulation in Table 2 is 0.37 oocysts person[-1] day[-1]. If this average were used as a single point estimate for exposure, then daily exposures to high doses of oocysts would never be predicted[8]. Indeed using this average exposure in a Poisson distribution would predict that three-fold the number of consumers are exposed to oocysts each day (compared to Table 2), but that none are exposed to more than about two or three oocysts in a day. The implications of ignoring the variation in pathogen exposures for risk assessment will depend on the infectivity of the pathogen under consideration and in particular the nature of the dose-response curve.

4 DOSE-RESPONSE CURVES FOR *C. PARVUM* - ESTIMATING THE RISK FROM
A SINGLE OOCYST

The infectivity of a calf strain of *C. parvum* has been determined in human feeding trials[10].
Healthy human adult volunteers were each given a dose of between 30 and 10^6 oocysts.
The proportion of volunteers infected at each dose is plotted in Figure 3.

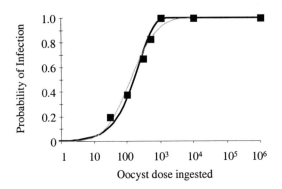

Figure 3   *Negative exponential dose response curve fitted by Haas[9] to C. parvum
human infectivity data of DuPont[10]. The log-probit curve ($\mu = 2.119$, $\sigma = 0.614$; log_{10}) is
shown for comparison (light line).*

The simulations in Table 2 suggest that even under outbreak conditions most drinking
water consumers are exposed to fewer than 30 oocysts. Indeed, half of those consumers
exposed ingest just a single oocyst (Table 2). Therefore the doses administered in the
human feeding trials[10] are generally too high for the application of modelling risks through
water. To estimate the risk from ingestion of low doses down to just a single *C. parvum*
oocyst, a process of low dose extrapolation of fitted mathematical dose response curves is
used. Haas fitted a negative exponential dose-response curve to the data[9]. This model
assumes that the oocysts act completely independently during infection and do not co-
operate with each other in overcoming the host defensive barriers. A second mathematical
model which could be used for a dose-response curve is the log-probit model[11]. This model
assumes that the oocysts do co-operate with each other during infection. It is also fitted in
Figure 3. The two models do not differ markedly over the range of doses administered in
the trial (30 to 10^6 oocysts). However, from Table 4 it is apparent that more considerable
differences exist on extrapolation to the low doses to which most drinking water consumers
are likely to be exposed (Table 2). Indeed the risk predicted by the log-probit curve from a
single oocyst is over 15-fold lower than that predicted by the negative exponential curve
(Table 4). This reflects that fact that if oocysts acted co-operatively then low doses (e.g.
one oocyst) would present greatly diminished risks. Experiments conducted with
salmonellae given to mice by mouth in the 1950s suggested that bacteria do indeed act
independently and not co-operatively[12]. It is generally accepted therefore that the negative
exponential dose-response curve (or in some cases the Beta-Poisson curve[1,11]) is

appropriate for MRA and that unlike certain potentially carcinogenic chemicals such as bromate[13], there is no threshold dose for micro-organisms through the oral route. Studies with gnotobiotic lambs concluded that the minimum infective dose is just one oocyst of *C. parvum*[14].

Table 4 *Risks from C. parvum predicted by low dose extrapolation of negative exponential model and the log-probit model (Figure 3).*

Dose Oocysts	Risk Negative Exponential (Independent action)	Log-Probit (Co-operative action)	Ratio
1	0.0041	0.00028	14.9
2	0.0083	0.0015	5.4
3	0.0125	0.0038	3.3
10	0.041	0.034	1.2

5 THE IMPORTANCE OF VARIATION IN PATHOGEN EXPOSURES IN RISK ASSESSMENT.

Table 5 compares the risks of *Cryptosporidium* infection predicted by the two dose-response curves for the exposures simulated in Table 2. The risk of infection predicted using the negative exponential dose-response curve is little affected whether the log-Normal variation in exposures is accommodated or whether just the arithmetic mean exposure is used. In contrast using the average exposure with the log-probit model under-estimated the risk by a factor of 5-fold. From Table 5, it would appear that the negative exponential dose response curve offers the best solution because it is independent of the spatial and temporal variation in oocyst densities within the treated water. It is concluded that modelling the variation in oocyst exposures in Table 2 is not important for MRA providing the negative exponential curve is appropriate for describing the dose-response relationship. Furthermore, it may be concluded from Table 5 that the risk of infection is directly related to the arithmetic mean oocyst density, which is in turn related to the total number of oocysts in the drinking water supply. Therefore any process which contributes to underestimating the arithmetic mean oocyst density in drinking water will also cause MRA to underestimate the risk of infection.

Table 5 *Predicted daily risks of Cryptosporidium infections (per 10,000 people) for Poisson-log-Normal distribution of oocyst exposures in Table 2 and dose response curves fitted in Figure 3. Risks predicted by just using the average exposure are presented for comparison.*

Dose-Response Curve	Distribution of Exposures Poisson-log-Normal	Arithmetic mean exposure (0.37 oocysts person^{-1} day^{-1}) in Poisson distribution
Negative Exponential	14.5	15.5
Log-Probit	9.3	1.8

5.1 Monitoring programmes tend to underestimate the total oocyst loading in drinking water and hence the risk of infection

From Table 3 it is apparent that 88% of 100-l samples contain fewer oocysts than the arithmetic mean of 30 oocysts. Further simulations showed that even averaging counts from ten 100-l samples underestimated the arithmetic mean over 70% of the time. This is because the most frequent count is zero and the all important high count samples are infrequent and likely to be missed except by the most intensive sampling programmes. Thus monitoring programmes based on taking a few 'spot' samples will in general underestimate the arithmetic mean pathogen exposure on the drinking water population and hence the risk. This may explain why there is a lack of a clear association between measured oocyst counts in drinking water and the observed illness in the population during waterborne outbreaks of cryptosporidiosis[4]. For this reason alone MRA models need to accommodate the variation in pathogen densities in treated waters.

6 THE NEED FOR MORE DOSE-RESPONSE DATA FOR WATERBORNE PATHOGENS

Most dose-response data available for waterborne pathogens were obtained from studies with healthy human adults[10]. Little or no data are available for the more susceptible individuals in the population such as children and the immunocompromised. Furthermore, MRA for *C. parvum* need to take into account acquired protective immunity[4]. The human infectivity data in Figure 3 were obtained for adult volunteers selected on the basis of having no serological evidence of past infection with *C. parvum*. Evidence on whether there is threshold dose is of critical importance for assessing the risks of BSE transmission to humans and cattle through environmental routes of exposure[2,3].

References

1. P. Gale, *Journal of Applied Bacteriology*, 1996, **81**, 403.

2. P. Gale, C. Young, G. Stanfield and D. Oakes, *Journal of Applied Microbiology, 1998,* **84**, 467.

3. P. Gale, *Letters in Applied Microbiology*, 1998, **27**, 239.

4. G. F. Craun, S. A. Hubbs, F. Frost, R. L. Calderon & S. H. Via, *Journal American Water Works Association*, 1998, **90(9)**, 81.

5. P. Gale, R. Lacey, G. Stanfield and D. Holt, *J Water SRT - Aqua*, 1997, **46**, 185

6. W. O. Pipes, P. Ward, S. H. Ahn, *Journal American Waterworks Association*, 1977, **69**, 664.

7. P. Gale, P. A. H. van Dijk and G. Stanfield, *J Water SRT - Aqua*, 1997, **46**, 117.

8. P. Gale, *Wat. Sci. Tech.*, 1998, **38(12)**, 7.

9. C. N. Haas, C. S. Crockett, J. B. Rose, C. P. Gerba, and A. M. Fazil, *Journal AWWA*, 1996, **88(9)**, 131.

10. H. L DuPont, C. L. Chappell, C. R. Sterling, P. C. Okhuysen, J. B. Rose & W. Jakubowski, *New England Journal of Medicine*, 1995, **332**, 855.

11. C.N. Haas, *American Journal of Epidemiology*, 1983, **118**, 573.

12. G.G. Meynell, *J. Gen. Microbiol.*, 1957, **16**, 396.

13. K. Chipman (This volume).

14. D.A. Blewett, S. E. Wright, D.P. Casemore, N.E. Booth & C.E. Jones, *Wat. Sci. Technol.*, 1993, **27(3)**,, 61.

Control of DBPs

THE USE OF ADVANCED WATER TREATMENT PROCESSES FOR THE REDUCTION OF BACKGROUND ORGANICS AND DISINFECTION BY-PRODUCTS

J. Shurrock, M. Bauer, M. Holmes and A. Rachwal

Thames Water Utilities Ltd,
Research and Technology,
Spencer House,
Manor Farm Road,
Reading, RG1 8DB

1 INTRODUCTION

Thames Water supplies 2700 MLd of drinking water to approximately 7.3 million customers. The majority of the water in the London region is supplied by five key water treatment works. In order to achieved compliance with EC Drinking Water Directive[1] and subsequent UK Water Supply (Water Quality) Regulations[2] the conventional water treatment process had to be improved. Key prescribed limits in the new regulations included an average concentration of 100 µg/l for total trihalomethanes and maximum concentration of 0.5 µg/l for total pesticides. Research has shown that conventional treatment comprising of primary and slow sand filtration prior to disinfection was unable to met these new standards particularly with respect to pesticides. In the United States Phase 1 of the Disinfection By-Products Rule (DBP) is proposing a limit of 80 µg/l for THMs.

The conventional water treatment process involved primary and slow sand filtration before disinfection. Many years research was undertaken within Thames Water to develop the advanced water treatment process (AWT) which would meet increasing customer demands as well as the new legislation.

The AWT process incorporated ozone and granular activated carbon (GAC) into the conventional process. Primary filtration stage at certain sites was upgraded/uprated by the construction of high rate dual media (sand/anthracite) rapid gravity filters (RGF). Pre-ozonation improved the RGF performance and reduced the particle loads to the slow sand filters.[3] In London the disinfection used chloramination, this ensured the residual in the distribution system was chloramines as opposed to free chlorine.

The patented GAC sandwich™ was implemented in Thames Water after pilot work demonstrated the efficiency of the process and development of new laser based methods for laying and removing a layer of GAC (Chemviron Carbon F400) in a slow sand filter. The GAC Sandwich essentially consisted of inserting a layer of GAC within a slow sand filter sandwiched between layers the of sand. The filter was operated in the same manner as a slow sand filter (SSF), except the GAC is periodically removed, regenerated and re-laid. The GAC Sandwich™ has the benefits of a biological slow sand filter as well as the adsorption capabilities of the GAC.

Extensive research has already documented the effectiveness of the AWT process for pesticide removal and subsequent compliance with the EC Directive.[4,5] The AWT process is capable of reducing background organic concentrations and preventing disinfection by-product formation. The majority of the research in Thames Water was undertaken at a demonstration scale plant at Kempton Park. The plant was designed so that the best process option for GAC and ozone in water treatment could be established. The plant was capable of delivering 5 MLd, this scale enabled a good comparison between the research site and the operational sites in London.

Research on the Kempton Park demonstration plant was approximately 1% of £350m spent on upgrading the London water treatment works. This paper looks at the success of the now operational AWT sites in London in meeting the EC drinking water standards.

2 SAMPLING AND ANALYSIS

Three operational sites in south west London were investigated for background organics and total trihalomethanes (TTHM). The operational sites and the demonstration plant were all fed from the same network of river Thames raw water filled reservoirs (typical retention time was 50 - 100 days).

2.1 Organics Removal - Ashford Common

Ashford Common water treatment works (WTW) capacity was 690 Mld. In 1993 the micro-strainers were replaced by dual media rapid gravity filtration. By mid 1996, 80% of the slow sand filters had been converted to GAC sandwich filters and multi stage ozonation had been commissioned. Ozone was dosed before and after RGF's.

2.1.1 Experimental Programme. In 1990 a trial GAC Sandwich™ filter and control slow sand filter were installed at Ashford Common. A variety of parameters were monitored to examine the background organics and pesticide removal from the experimental and control beds. The sampling continued intermittently for several years.

Apparent colour, total organic carbon (TOC) and absorbance at uv254 nm were selected as key indicators of background organics. Percentage removals of each parameter have been examined to illustrate the effectiveness of the GAC Sandwich. When ozone was commissioned in 1996 the inlet monitoring programme was continued to assess the impact of ozone in reducing concentrations of background organics.

2.2 Trihalomethane Removal - Hampton and Kempton Park

2.2.1 Hampton. Hampton WTWs capacity was 790 Mld. The process comprised of sand RGFs followed by 20% slow sand filters and 80% GAC sandwich filtration. In early 1998 multistage ozonation will be commissioned.

2.2.2 Kempton Park. Kempton Park's WTWs capacity was 200 MLd. This site was similar to Hampton with two exceptions. Secondary filtration occurred only through GAC sandwiches and ozonation after the RGF's had been installed.

2.2.3. Experimental Programme. Trihalomethanes were sampled for from the main tap (after disinfection) on a monthly basis. Individual and total trihalomethanes were analysed.

3 RESULTS AND DISCUSSION

Research showed that advanced water treatment incorporating GAC and ozone was highly effective at pesticide removal.[4,5] Another benefit of this treatment was the removal of background organics and the limitation of disinfection by-product formation.[6]

3.1 Organics Removal

3.1.1 Impact of GAC Sandwich™. The trials at Ashford Common examining organics removal show a clear difference between the experimental GAC sandwich and the control slow sand filter. For TOC, apparent colour and uv254 nm absorbance the GAC sandwich showed a greater percentage removal than the conventional slow sand filter (see Figures 1, 2 and Table 1). This confirmed the research results from the demonstration plant at Kempton.[6]

The removal of all three parameters by the GAC Sandwich was higher than average in the first six months. (Table 1). Virgin GAC is known to have higher adsorption capabilities than older GAC (older than 6 months). After this initial period the adsorption sites on the GAC become depleted and the GAC becomes biologically active. The biological activity may facilitate the long term advantage of the GAC sandwich over the conventional SSF. Even after six years of operation the GAC sandwich still removed more background organics than the conventional SSF.

The smallest advantage of the GAC Sandwich over the SSF was seen by TOC removals (Figure 3). The percentage removal rates in the first six months are not

Table 1: *Percentage Removals for Organics Parameters at Ashford Common.*

	TOC		UV 254nm		Colour	
	GAC	SSF	GAC	SSF	GAC	SSF
Average	39	21	55	15	77	33
First six months' average	56	20	80	15	92	31

available but in the second six months removal rates were 56%, this declined to an average of 39% for the remainder of the six years. The GAC Sandwich achieved 18% more TOC removal than the traditional SSF even after 6 years of operation.

This data shows the GAC Sandwich combined the benefits of biological slow sand filtration and GAC adsorbance capabilities. It is important to note even after six years of operation the GAC sandwich still has a distinct advantage over the conventional SSF by producing a higher quality filtrate with fewer organics present. Additionally the overall operating and capital expenditure costs of the GAC Sandwich was less than for the combination of GAC adsorbers and slow sand filters.[7]

It is well understood that the concentrations of disinfection by-products (DBP's) and TTHM's are dependant on the organics present in the water as well and the type and degree of disinfection utilised. A linear correlation is found between trihalomethane formation potential and organic parameters such as colour or TOC, as a result they are considered surrogates for THM formation potential.[8] Therefore by maximising the organics removal through the plant and optimising the disinfection process in terms of applied dose and contact time together with the use of chloramination in distribution the risk of DBP occurring was also minimised.[9,10]

3.1.2. Impact of Ozonation. In the last two years of these trials ozone was used at Ashford Common before and after the RGF's. Ozone is known to help breakdown complex organics which are then removed by further biological filtration stages.[8,11]

Adsorbance at uv254nm detects the C=C bonds present in complex organics. Before ozone was commissioned the average uv254nm concentrations reaching the GAC Sandwich filter and SSF were 11.2 abs/m. After ozone was commissioned the UV 254nm concentrations fell to 8.0 abs/m. The decline in uv254nm concentrations illustrates the effectiveness of ozone for the breakdown of complex organics and subsequently limit the disinfection by-product formation.[12]

Figure 1, *Percentage Removal of Adsorbance at 254nm at Ashford Common.*

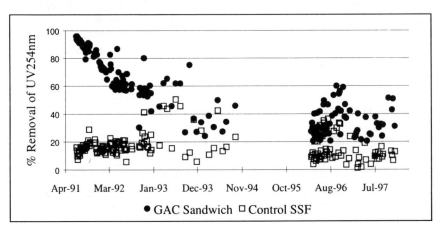

Figure 2, *Percentage Removal of Apparent Colour at Ashford Common.*

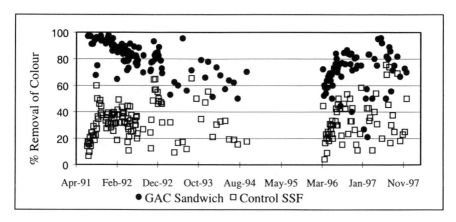

Figure 3, *Percentage Removal of Total Organic Carbon at Ashford Common.*

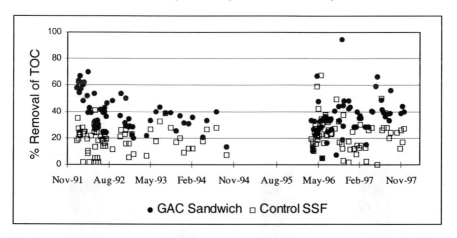

3.2 Trihalomethane Removal

There is a clear difference between the two sites examined for THM concentrations leaving the works (Figure 4). Hampton water treatment works showed consistently higher concentrations of TTHM's than Kempton Park, although both sites were well within the 100 µg/l statutory limit.

The Kempton Park process included ozone and 100% GAC coverage whereas Hampton had 80% GAC coverage only. This demonstrated the superior treatment currently available at Kempton Park. The inclusion of ozone increased the organics removal through the process and subsequently prevented TTHM formation. As ozone is commissioned at Hampton the concentrations of TTHM's are expected to fall to the same concentrations found at Kempton Park.

Figure 4, *Concentrations of Total Trihalomethanes at Two London Works.*

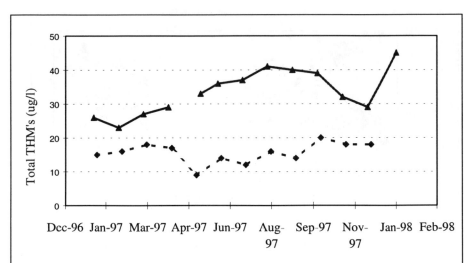

Figure 5, *A Comparison of THM Species for Hampton and Kempton Park.*

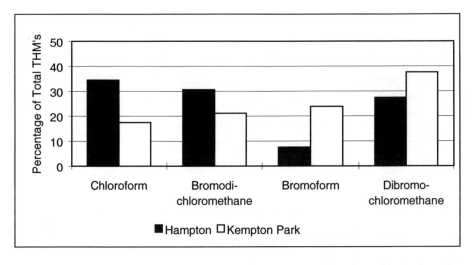

As these TTHM concentrations were measured shortly after chlorination the concentrations are expected to be slightly lower than those found in the distribution system. The use of chloramines will substantially limit the amount of free chlorine in the distribution system, in the absence of free chlorine THM's should not form.

3.3 Trihalomethane Speciation

Ozone influenced the individual THM species which occurred (Figure 5).[13] As expected Hampton works had higher levels of chloroform and bromodichloromethane than Kempton Park. The brominated THM species are more common at Kempton Park as a result of the ozonation.

When ozonation is commissioned at the Hampton works the prevalent species of THM's is expected to switch to the brominated species. This will reduce the concentrations of bromodichloromethane which has the highest health risk.[15] As a result of the higher health risk it is also possible that the limit for bromodichloromethane will be lower than other THM species in the future.[15]

4 CONCLUSIONS

1. The GAC sandwich has proved to be more effective at organics removal than conventional slow sand filters even after six years of operation.

2. Ozonation has been show to reduce concentrations of complex organics by forming simpler compounds which are removed by a further biological step. In turn this limits the formation of total trihalomethanes.

3. Ozonation results in the formation of predominately brominated THM's as opposed to chlorinated THM's at more conventional WTW. Dichlorobromomethane has the highest health risk, but its formation is more limited after ozonation.

4. The combination of ozone and GAC sandwich in the AWT process is effective at meeting customer demands and legislation by reducing pesticide, organics and TTHM concentrations.

ACKNOWLEDGEMENTS

The authors wish to thank Thames Water Engineering Director, Steve Walker for permission to publish the paper. The authors would like to acknowledge the contributions to this work from the Thames Water operational staff, scientists and researchers. The views expressed are those of the authors.

REFERENCES

1. Council for European Communities. 80/778/EEC. Official Journal, 1980.
2. Water Supply (Water Quality) Regulations, HMSO, 1989.
3. M. Chipps, M. Bauer, A. Delanoue, A. Rachwal. Proc. 11th Ozone World Congress, San Francisco, 1993, **1**.
4. B. Croll, Ozone Science and Engineering, 1996, **18**, 19.
5. D. Foster, M. Holmes and A. Rachwal, Proc. International Ozone Association European Group, Zurich.1994.

6. A. Rachwal, D. Foster, M. Holmes and M. Hillyer, Proc. 11th Ozone World Congress, San Francisco, 1993, **2**.

7. M. Bauer, B. Buchanan, J. Colbourne, D. Foster, N. Goodman, A. Kay, A. Rachwal, T. Sanders, Proc. Specialised Conference of Advanced Water and Integrated Water System Management into the 21st Century, Osaka, Japan 1994.

8. C. Cable and R. Jones, 'Advances in Slow Sand and Alternative Biological Filtration' Wiley and Sons, 1996, p 29.

9. P. Chen and G. Rest, Public Works, January 1996.

10. L. Cipparone, A. Diehl, G. Speital, JAWWA, 1997 **89** (2), 84.

11. S. Krasner, M. Sclimenti, B. Coffey,. 1993. JAWWA, 1993 **85** (5), 62.

12. A. Orlandini, A. Siebel, A. Graveland, J. Schippers, Proc. International Ozone Association European Group, Zurich 1994.

13. A. Miltner, H. Shukariy, R. Summers, 1992. JAWWA, 1992, **84**(11) 53.

14. World Health Organisation, Second Edition, 1993, **1,** Recommendations Geneva.

15. S. Krasner, M. Sclimenti, E. Means, J. Symonds, L. Simms, Proc. 1992 AWWA WQTC, Toronto, **2**, 1745.

TREATMENT STRATEGIES FOR DISINFECTION-RELATED TRIHALOMETHANE FORMATION: IMPACT OF COLOUR IN PUBLIC WATER SUPPLIES TO KLAIPEDA, LITHUANIA

Scott, P. A. R. S., Dean, C., Jones, M. A.
Scott Wilson, Basingstoke, UK

Sulga, V.
Vilnius Technical University, Vilnius, Lithuania

Klimas, A.
Vilniaus Hidrogeologija, Vilnius, Lithuania

1 INTRODUCTION

The city of Klaipeda, a major port on the Lithuanian coast, depends on water supplied by three wellfields. In 1991 water supplied by the largest of these wellfields, Wellfield No.3, rose to a peak of about 65,000 m^3 day^{-1}, but declined steadily following Independence to about 40,000 m^3 day^{-1} in 1996. Over the same period the quality of water supplied by Wellfield No.3 deteriorated progressively, with colour levels in treated water exceeding Lithuanian standards for potable water.

To improve the quality of water supplied by Wellfield No.3, and achieve a yield of 60,000 m^3 day^{-1}, a study was carried out to determine the most feasible solution. This addressed the high colour and, more importantly, the formation of trihalomethanes (THM) which occurs on disinfection of these highly coloured waters. A diverse range of solutions was considered. They included conventional physico-chemical treatment, but perhaps of greater interest was the more novel approach of using natural aquifer filtration for reduction of colour, and its significance for limiting THM production.

1.1 Hydrological Setting and Operation of Wellfield No.3

Wellfield No.3 is sited on the banks of the Kaiser Wilhelm Canal (KWC) which is crucial to the operation of the wellfield. Although groundwater is abstracted from a shallow, unconfined sand aquifer, artificial recharge of this aquifer is essential to enable the wellfield to meet its supply targets; the source of water used to artificially recharge the aquifer is the KWC. The importance of artificial recharge is illustrated well by the estimate that only 10 % of the water abstracted is natural groundwater, the remainder being derived from the KWC, either as artificial recharge or as direct flow from the canal into the aquifer. Consequently the KWC plays a fundamental role, controlling both the quantity of water available for abstraction as well as its quality.

The key elements of the wellfield operating system are a series of, (a) recharge channels, each 3.1 km long, through which the aquifer is recharged, and (b) infiltration galleries, with a total length of 3 km, plus 63 wells which abstract groundwater (Figure 1). Water flows from the recharge channels to the abstraction galleries and wells, the rationale being that water derived from the KWC is treated by natural, in situ filtration within the aquifer before it is abstracted.

Figure 1 *Layout of Wellfield No.3*

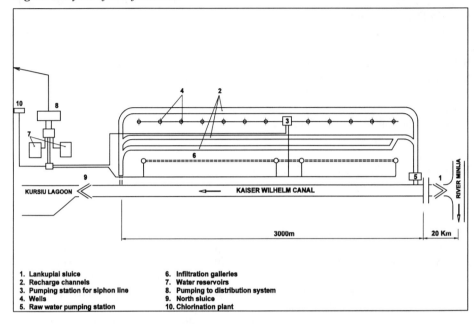

1. Lankupiai sluice
2. Recharge channels
3. Pumping station for siphon line
4. Wells
5. Raw water pumping station
6. Infiltration galleries
7. Water reservoirs
8. Pumping to distribution system
9. North sluice
10. Chlorination plant

1.2 Water Quality and Current Treatment Method

The water pumped from the KWC, derived originally from the Minija River, is affected during its passage along the 20 km long canal by inflows of organic-rich water from adjacent swamps, with a resultant deterioration in quality from 1991 to 1996 (Table 1[1,2]). Even with natural filtration within the aquifer the quality of the abstracted water, especially its colour, has also deteriorated (Table 2[1,2]); since 1993 the minimum colour levels measured have exceeded key limits set by the existing and proposed Lithuanian potable water standards. These standards are illustrated as follows:

a) water classified as 'satisfactory' by the existing standards has a maximum permissible colour (MPC) of $20°$ on the chromium-cobalt scale (abbreviated here as $20°$ Cr-Co);

b) water classified as 'satisfactory' by the proposed standards has an MPC of $25°$ Cr-Co, with 'good' water having the same magnitude (i.e. 20) as the MPC set by European potable water standards.

The current treatment method for the abstracted water has remained unchanged during the period of quality deterioration, comprising only disinfection by simple chlorination. Using liquid chlorine at a dose of 3 to 4 mg l[-1], residual chlorine levels do not exceed 0.3 to 0.5 mg l[-1] in the supplied water. Following disinfection the water is pumped directly into the city water distribution system. However, the reaction of chlorine with natural dissolved organic matter in the water produces odours derived from THMs such as chloroform. In previous studies in 1989[3], chloroform was measured at concentrations of up to 68 µg l[-1] in treated water supplied from the wellfield. Subsequent investigations in this study in 1996 have shown that several THMs can be produced (Table 3), but in line with other studies[4] their relative magnitude, in order of decreasing concentration was as follows, chloroform>bromodichloromethane>chlorodibromomethane>bromoform.

Table 1 *Colour Values for Raw Water Pumped from the KWC*[1,2]

Year	Colour/degrees Cr-Co scale		
	Maximum	*Minimum*	*Average*
1991	140	43	79
1992	128	27	71
1993	196	51	96
1994	65	42	52
1995	248	70	122
1996	200	184	192

Table 2 *Colour Values for Treated Water Abstracted from Wellfield No.3*[1,2]

Year	Colour/degrees Cr-Co scale		
	Maximum	*Minimum*	*Average*
1991	29	19	25
1992	32	19	26
1993	40	25	32
1994	40	22	30
1995	36	28	33
1996	35	34	34.5

Table 3 illustrates that THM concentrations are very low in the raw water feed to Wellfield No.3, i.e. the KWC. On chlorinating the water abstracted from the wellfield total THM concentrations of 57 µg l^{-1} were produced, comparable to those measured in 1989[3]. Disinfection was achieved in these water samples by manually adding a hypochlorite solution to produce a chlorine dose of 5 mg l^{-1}, instead of using the installed chlorination system which doses at concentrations of 3 to 4 mg l^{-1}.

There were difficulties in analysing for THMs in Lithuania, and the analyses for those samples summarised in Table 3 were carried out in the UK. This, at least partially, accounts for the discrepancy between the high minimum colour values quoted for 1996 in Tables 1 and 2, and those colour values quoted in Table 3. The former two sets of values were measured in Lithuania using the Soviet standard based on potassium dichromate and hydrated cobalt sulphate[5], while the latter set were measured in the UK using the Blue Book method based on potassium chloroplatinate and hydrated cobalt chloride[6]. The situation was complicated further as during the study, Lithuanian laboratories switched to a colour determination method based on the ISO method[7]. A method comparison completed in the UK showed large differences between the results of the Soviet and Blue Book methods, but the ISO and Soviet methods produced similar results.

There was no limit set for total THMs in either the existing Lithuanian standard for potable water or the proposed standard, but the latter set limits of 30 and 25 µg l^{-1} for chloroform and bromodichloromethane respectively. It can be seen from Table 3 that the concentrations of chloroform are at, or exceed permissible concentrations in the abstracted water samples when disinfected using chlorination, while concentrations of bromodichloromethane are about 20 % of the permissible concentration in treated waters.

Table 3 *Colour and Trihalomethane Concentrations in Water Abstracted from Wellfield No.3 following Chlorination and in Water from the KWC*

Water Source	Colour /mgPt l^{-1}	Trihalomethane concentration/μg l^{-1}				
		Total	Chloro-form	Bromodichloro-methane	Chlorodibromo-methane	Bromo-form
2/4/96						
KWC	19	0.7	<0.7	<0.5	0.7	<0.5
Treated	16	47.3	30	5.3	12	<0.5
28/6/1996						
KWC	45	2.2	<0.7	<0.5	<0.5	<0.5
Treated	21	57	52	5	<0.5	<0.5

2 TREATMENT STRATEGIES

To address the problem of excessive colour levels and associated production of THMs in the chlorinated water, several treatment strategies were investigated. These included, a) enhancing the natural filtration characteristics of the aquifer, and b) physico-chemical treatment, with several variations on these two themes.

Two other options were considered, neither of which involve natural or artificial treatment of the water. The first was replacement of Wellfield No.3 with a more distant wellfield and supply pipeline, but the estimated construction cost of £ 13.5 million[8] was far in excess of the costs for other options (see Section 2.4). The second was to improve operation of the KWC and to eliminate inflows of organic-rich water from the adjacent swamps. However, only minor modifications to the canal would be possible without affecting water levels and flows in the swamps, which are of designated ecological importance, such that water treatment would still be necessary to reduce colour levels and limit THM production. Consequently, neither of these options was considered further.

2.1 Treatment Options Considered

To limit production of THMs, disinfection by chloramination rather than chlorination was considered briefly, but this would not have reduced colour levels to the required potable standards. Thus, reduction of THM concentrations was addressed by seeking to lower colour, and consequently reduce the potential for THM production. The feasibility of implementing various treatment strategies considered the following options in detail:
a) enhanced aquifer filtration, building on the natural processes currently operating in Wellfield No.3 by,
 - increasing the distance between the raw water feed and the abstraction points, and
 - injecting abstracted water to allow recirculation, i.e. two stage filtration;
a) physico-chemical treatment, taking a more conventional approach to promote coagulation and flocculation of the organic matter producing the high colours, by treatment of,
 - raw water from the KWC prior to artificial recharge, termed pre-treatment, and
 - abstracted water following natural aquifer treatment, referred to as post-treatment.

2.2 Enhanced Aquifer Filtration

2.2.1 Lengthened Filtration Path. Over a period of 2.5 months, field investigations of natural aquifer filtration were carried out by monitoring the colour of water pumped from one of the infiltration galleries 47.5 m from the KWC. This infiltration gallery was not part of the active abstraction system and initially the colour of groundwater, as measured in a series of monitoring wells, was about 40° to 60° Cr-Co. However, thereafter water quality deteriorated, with water abstracted from the infiltration gallery after 2 months having a colour of almost 100° Cr-Co (Figure 2). The results show that aquifer filtration occurs mainly within 25-30 m from the KWC, with colour removal decreasing significantly at greater distances. Despite large colour reductions, and thus the potential for THM production, the colour of abstracted water exceeded the existing MPC, with there being little indication that a lengthened filtration path would improve colour removal.

2.2.2 Two Stage Filtration. Two month long field trials investigated the feasibility of abstracting water from an infiltration gallery, then injecting it back into the aquifer for a second stage of colour removal. Final abstraction was from a second infiltration gallery 42 m from the recharge channel. The results were complicated by increases in the colour of the water injected into the aquifer during the trial, but it was clear that the extra stage of treatment further reduced colour. However, the final colour of the abstracted water still exceeded the existing MPC, indicating that two stage filtration can reduce colour levels, and thus the potential for THM production, but not to acceptable levels for potable water.

Although aquifer filtration has the merits of not needing treatment additives and being cheap, its performance was not promising. In fact the aquifer's capacity to continue lowering colour over the proposed 40 year wellfield life was noted as an issue of concern.

Figure 2 *Water colour measured in natural aquifer filtration trials with best fit curves*

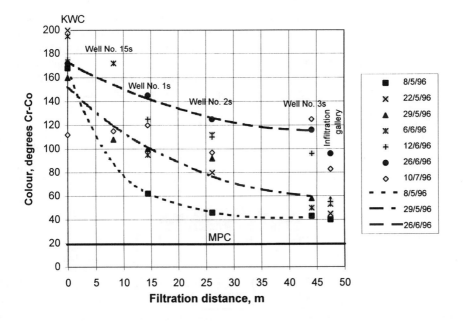

2.3 Physico-Chemical Treatment

Bench scale testing was carried out using an active aluminium oxide dose ranging from 10 to 40 mg l^{-1}; the aluminium oxychloride coagulant used contains about 9 % active aluminium oxide. For water abstracted from the wellfield having a colour of 35 mg Pt l^{-1}, measured using ISO 7887[7], bench tests showed that colour could be reduced to 9 mg Pt l^{-1} with a dose of 20 mg l^{-1} aluminium oxide. By adding a 0.8 mg l^{-1} polyacrylamide flocculant (Magnafloc LT22), a colour of 11 mg Pt l^{-1} could be achieved with an aluminium oxide dose of only 10 mg l^{-1}. However the large flocculant dose, exceeding the proposed Lithuanian standard of 0.5 mg l^{-1} for polyacrylamide dosing, and high cost make its continuous addition technically and economically impractical. Optimal doses of 19.4 mg l^{-1} aluminium oxide and 0.3mg l^{-1} polyacrylamide were identified. Apart from aluminium oxychloride, two other coagulants were considered but neither were suitable; aluminium sulphate is less active, less easy to use in its solid form and is more expensive, while Magnafloc LT31, a cationic polyelectrolyte, was not effective in removing colour.

Using aluminium oxychloride, two treatment options were considered. Firstly, treatment after abstraction from the wellfield (i.e. post-treatment), and secondly, treatment of raw water from the KWC before introducing it as artificial recharge (i.e. pre-treatment). However, there were concerns over the technical feasibility of the latter including, a) the potential removal of organic matter accumulated within the aquifer by treated water having a much lower colour, and the associated increase in potential for THM production, and b) the impact of residual aluminium on aquifer permeability and water quality. The risk of these hazards occurring put the feasibility of the pre-treatment option in doubt.

Pilot testing of aluminium oxychloride treatment, in a system including an upflow filter to remove turbidity resulting from coagulation (Figure 3), illustrated that this process was successful in removing sufficient colour to limit production of THMs on chlorination of the treated water (Table 4). It can be seen from Table 4 that the colour does not achieve the 'satisfactory' level of the existing Lithuanian standard, and thus, the 'good' level of the proposed standard, in samples treated on successive days. This is because the dose of aluminium oxide added to the pilot system was well below the optimum dose, which prevented effective coagulation and flocculation of the organic matter. Despite this incomplete removal of colour, concentrations of chloroform and bromodichloromethane produced on chlorination were well below the proposed Lithuanian standards.

Figure 3 *Schematic flow diagram of water treatment system*

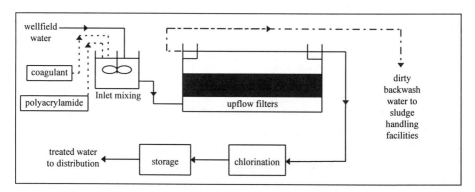

Table 4 *Trihalomethane Production in Water Treated using Aluminium Oxychloride*

Water Source (date)	Colour /mgPt l⁻¹	*Trihalomethane concentration/µg l⁻¹*			
		Chloro-form	*Bromodichloro-methane*	*Chlorodibromo-methane*	*Bromo-form*
Untreated wellfield water	52	64	9.3	1.1	<0.5
Pilot plant treated water; 1	-	11	1.6	0.5	<0.5
Pilot plant treated water; 2	23	14	2.1	0.7	<0.5

2.4 Economic Analysis

The treatment options were evaluated economically by calculating the Unit Water Cost (UWC), i.e. the price needing to be charged per unit of water produced throughout the life of the scheme such that the present value of revenue would exactly equal the present value of costs. Assuming the wellfield produced 60,000 m³ of water daily for a period of 40 years, the economics of producing potable water were evaluated by considering the marginal costs of new treatment works only. The following additional assumptions were made, a) replacement costs of pumping, generator and treatment plant after a useful life of 20 years, at a sum equal to half the initial plant cost, b) annual maintenance costs of 1.1 % of the total construction cost.

The results show that capital costs and UWC for physico-chemical treatment are far less than constructing a new wellfield outside Klaipeda. Furthermore, it appears that construction costs of the pre-treatment option are significantly lower than the costs for the post-treatment option. Thus, despite its technical problems, physico-chemical treatment prior to artificial recharge appears more favourable. Nevertheless the marginal net present value of the UWC for post-treatment is 2.2 pence m⁻³, as opposed to 2.5 pence m⁻³ for pre-treatment, indicating both technically and economically that physico-chemical post-treatment for removal of colour and limitation of THM production is the favoured option.

Table 5 *Comparison of the Unit Cost of Water for the Selected Treatment Options*

Economic Factors	*Post-Treatment*	*Pre-Treatment*	*New, Distant Wellfield*
Capital cost (£)	1,600,000	1,300,000	13,500,000
Unit Cost of Water (pence m⁻³)	2.2	2.5	6.3

3 DESIGN OF SELECTED TREATMENT WORKS

The post-treatment option, as outlined schematically in Figure 3, was selected for implementation and is currently under design. The design aims to meet the proposed Lithuanian potable water standards for colour (20⁰ Cr-Co), chloroform (30 µg l⁻¹) and bromodichloromethane (25 µg l⁻¹). The new facilities comprise elevated inlet mixing chambers, incorporating flash mixing of the water treatment chemicals, followed by a short period of mixing in flocculation tanks. These are followed by a 2.8 m deep, rapid gravity upflow filter in which secondary particulate flocculation and subsequent solids removal takes place. The water is then chlorinated via the existing chlorination facilities and stored in reservoirs prior to distribution.

The optimal chemical dosing rates of 19.4 mg l^{-1} aluminium oxychloride coagulant, measured as aluminium oxide, and 0.3mg l^{-1} polyacrylamide flocculant have been used as a basis for design. However the dosing facilities will allow for significant variation in dosing rates to enable flexibility of operation and on-site dosing optimisation.

4 CONCLUSIONS

To meet the proposed Lithuanian potable standards in water supplied from Wellfield no. 3 in Klaipeda, a revised treatment process was required to remove high levels of colour and limit the formation of THMs on chlorination. The philosophy adopted to provide potable water was to treat the colour problem rather than addressing THM production directly.

Investigations of both conventional physico-chemical treatment and the more novel enhancement of natural aquifer filtration showed that the latter, although cheaper and preferable as no treatment additives are required, did not perform well in field trials. In fact concerns were raised over the aquifer's capacity to continue removing organic matter, and thus lower colour, during the proposed wellfield lifetime. Trials showed that treating abstracted groundwater by coagulation/flocculation using aluminium oxychloride and polyacrylamide was most cost effective, meeting proposed potable standards for colour and allowing continued use of chlorine disinfection without formation of THMs.

Acknowledgements

This World Bank funded study was carried out for JSC Klaipeda Water. The assistance of JSC Klaipeda Water staff and permission to publish this paper is gratefully acknowledged.

References

1. Klaipeda Environment Project, 'Feasibility Study Final Report', 1994.
2. Scott Wilson CDM in association with Vilniaus Hidrogeologija, 'Klaipeda Wellfield No.3 Development Study, Final Report', 1996.
3. A. Kondratas et al, 'Ecological and hydrogeological studies of Klaipeda Wellfield No.3 and selection of measures of optimisation', 1989 (Lithuanian).
4. F. W. Pontius (editor), 'Water Quality and Treatment: A Handbook of Community Water Supplies' 4th edition, American Water Works Association, McGraw-Hill, 1990.
5. GOST 3351-74, 'Test methods for taste, smell, colour and turbidity', 1974 (Russian).
6. Anon., 'Colour and turbidity of waters: Methods for the examination of waters and associated materials', HMSO, London, 1981.
7. International Standard ISO 7887, 'Water quality - Examination and determination of colour', 1985.
8. A. Klimas et al, 'Hydrogeological evaluation of alternative water supply sources for Klaipeda', 1994 (Lithuanian).

APPLYING PUBLISHED BROMATE FORMATION DATA TO OZONE CONTACTOR DESIGN

A. F. J. La Trobe-Bateman

Montgomery Watson
Terriers House
201, Amersham Road
High Wycombe

1 INTRODUCTION

The United Kingdom water supply industry is a regulated business. In addition to the legal requirement to supply water that is "Wholesome" according to the definitions in Part II of the 1989 Water Supply, Water (Water Quality) Regulations, the water supply companies are also required to satisfy requests for information from the Drinking Water Inspectorate (DWI). These requests are sometimes of a "what-if" nature, and can come at a time when there is little information available to answer the questions. A model of the formation or removal of the parameter in question enables what-if questions to be answered quickly, and with a minimal cost to the industry, and consequently the customer. The accuracy of early answers is restricted because initial modeling is based on limited input data, and uncertainty as to the governing parameters. The work leading to this submission was driven by a water company needing to predict the bromate formation at existing ozonation plant, and at several other plants under construction. The DWI were interested in the cost of achieving different levels of compliance with different maximum permissible concentrations. To be of any use, the work had to be able to use the historical data available at the time.

The latest version of the EU Revised Drinking Water Directive issued on 16 October 1997 has specified a parametric value of 10 μg/L for bromate. Phased compliance is required, at 25 μg/L within 5 years, and 10 μg/L within 10 years of the Directive coming into force.

2 FACTORS EFFECTING BROMATE FORMATION

Bromate formation in water treatment has been found to be affected principally by the following water quality parameters:

2.1 Bromide Level

The effect of bromide level on the formation of bromate in ozonated water is demonstrated by Shukiary, Miltner and Summers[1]. They applied the same ozone doses to samples of the same water, but with different bromide spikes. Their actual aim was to investigate the effect of bromide on total trihalomethane formation, but they also measured the resulting bromate levels. In each case bromates started to form at a transferred O_3/DOC (dissolved organic carbon) level of approximately 0.5 mg/mg. The initial bromide levels three samples were 51.7, 258, and 550 μg/L, which is in the ratios 1:5.1:10.8. At the maximum O_3/DOC dose of 2.54 mg/mg the bromate levels formed in each sample were in the ratios 1:4.9:7.4. This shows a strong correlation between bromide concentration and bromate formation.

A similar pattern is shown by the results of Amy, Siddiqui, Ozekin and Westerhoff[2] published in 1995. In cases where the ozone dose was sufficient to produce bromate, the amount produced was linearly related to the initial bromide concentration.

2.2 Dissolved Organic Carbon (DOC)

DOC is a major constituent of the ozone demand of a natural water. Ozone reacts with this before it tends to form bromates from bromide. It has been observed that a transferred ozone dose of at least 0.5 mg/mg DOC is required before bromates are formed at all, although for some waters this value can be as low as 0.29 (e.g. reservoir softened Biesbosch water)[3].

2.3 Ozone Residual

An ozone residual can only develop after the ozone demand of the water has been satisfied. Ozone demand can be thought of as compounds in the water whose reaction with ozone is almost instantaneous, e.g. DOC above. The mechanisms in the formation of bromate require the presence of an ozone residual in both intermediate steps, as well as the final step from hypobromite to bromate in the molecular mechanism, and bromite to bromate in the radical mechanism.[4]

2.4 Ozone Contact Time

The overall reaction to form bromate takes minutes rather than seconds. Amy, Siddiqui, Ozekin and Westerhoff show that the reaction is 50-75% complete in 10 minutes[2].

The process requirement for bromate formation is ozone contact time (c.t)[4]. The larger the c.t the greater will be the formation of bromates. An ozonation chamber which is designed in such a way as to minimize the ozone residual at any time will produce only low levels of bromate.

2.5 pH

The molecular mechanism for the formation of bromate from bromide involves the initial oxidation of bromide, Br^-, to the hypobromite ion, OBr^-. At low pH OBr^- reacts with hydrogen ions to form undissociated hypobromous acid, $HOBr$. This effectively removes the intermediate species from the reaction mechanism and inhibits the formation of bromate.

The radical pathway for the formation of bromate from bromide does not involve the hypobromite ion at all, however the formation of the OH⁻ radicals that are required for this mechanism is also inhibited by low pH. (c.f. the fact that advanced oxidation, which relies on radical formation, is carried out at high pH, at least above 7.5.)

2.6 Hydrogen Peroxide Dosing

The effect of hydrogen peroxide dosing on bromate formation during ozonation depends strongly on the H_2O_2/O_3 dose ratio. At low H_2O_2/O_3 doses bromate formation is increased for a given O_3 dose. At higher H_2O_2/O_3 dose ratios bromate formation is decreased for a given O_3 dose. This is thought to be because low H_2O_2 doses increase the concentration of hydroxyl radicals in the system and so produce more bromate via the radical mechanism. However at higher dose ratios the H_2O_2 reacts directly with the ozone and inhibits both the formation of radicals and the oxidation of bromide to bromate via the molecular reaction mechanism.

2.7 Ammoniacal Nitrogen

Hypobromite reacts with ammonia to form monobromamine or dibromamine, depending on the $HOBr/NH_4^+$ ratio. These are slowly oxidised back to intermediates in the bromate formation mechanism. The overall effect of ammonia is to delay the formation of bromate.[4]

2.8 Temperature

Bromate formation is increased by the temperature at which ozone dosing is carried out. A weaker positive temperature correlation is also apparent if the contact time is also carried out at higher temperatures.[5]

2.9 Alkalinity

Inorganic carbon plays no part in the molecular mechanism for bromate formation, but can scavenge OH⁻ radicals to form $CO_3^{\bullet-}$ radicals which themselves take part in the formation of bromate[4]. The net effect of increased inorganic carbon is to increase bromate formation for a given water[6].

3 PUBLISHED BROMATE FORMATION MODELS

There are two approaches to reaction modeling. The first is to identify all relevant reaction mechanisms and establish the rate constants. The second is to collect empirical data from as many reactions as possible, identify the governing parameters and formulate empirical equations that reproduce the same results. Both approaches can lead to models that are calibrated, but the first requires detailed understanding of the science and therefore can take longer to develop sufficiently to be useful in engineering design. The empirical approach can predict relatively accurately and quickly, as long as the input data for prediction lie within the bounds of the data used to generate the model. At the time this work was carried out, no empirical bromate formation models had been identified in the literature.

Subsequent to the development of the simple model used in this paper, Song, Minear, Westerhoff and Amy[6] have published an empirical model that includes all the parameters identified above.   They have reported that increasing bromate formation results from increasing the following parameters, in order of significance: pH>O_3 dose>Br‾>Alkalinity. Bromate formation is decreased by, again in order of significance: DOC>NO_3-N.

4 DEVELOPMENT OF SIMPLE BROMATE FORMATION MODEL

The model used in this case relates bromate formation to bromide concentration as a function of O_3/DOC.

Linear approximations in terms of O_3/DOC dose ratio and raw water bromide level were derived for four surface waters based on reported results of bromate formation.   The formulae reproduce worst case of the reported values.

The expected water quality at the relevant water treatment sites was substituted in each formula and the results plotted on a graph showing computed bromate formation versus ozone dose.   The predicted bromate formation from each of the models is different, as shown in figure   1 below, and the worst predicted case is taken for the purposes of estimating compliance to a specified limit.

Figure 1 - *Bromate prediction model with pilot test results superimposed.*

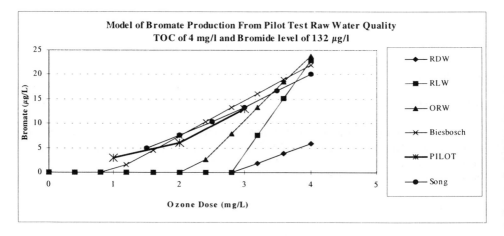

Key	Water Source	DOC mg/L	Bromide μg/L	pH
RDW[3]	Pre-treated Rhine water after dune infiltration	2	202	n/a
RLW[3]	Pre-treated Rhine water from IJssel lake	4	107	n/a
ORW[1]	Ohio River Water	2.2	51	7.65
Biesboch[3]	Reservoir softened water	2.4	121	7.8

4.1 Calibration of the Bromate Formation Model

Bromate formation results from a pilot test have been compared with model predictions for the same input water quality. The comparison is also given in figure 1. A prediction by the Song *et. al.* model for the same DOC, bromide and pH as the pilot plant included in the figure shows that the early model gave reasonable results for ozone doses between 1.5 and 4 mg/L.

4.2 Applying the Model to an Existing Ozonation Process

Bromate formation results from Beisbosch water most closely fit the treatment works pilot results. On the assumption that bromate formation at the full scale plant will be similar to the Biesbosch, the bromate production in table 1 following was predicted.

Table 1 *Bromate Production at Existing Treatment Works, based on Biesbosch Bromate Formation Data*

	Worst Case			Average		
Pre-ozone Dose	2	mg/L		1		mg/L
Main Ozone Dose	3	mg/L		1.5		mg/L
Bromide Level	174	μg/L		132		μg/L
	DOC mg/L	Ozone dose mg/L	Bromate μg/L	DOC mg/L	Ozone dose mg/L	Bromate μg/L
Pre-Ozone Contactor	5.1	2	5.6	6.3	1	0.0
Main Ozone Contactor 1	3	1.2	5.9	3.5	0.6	0.0
Main Ozone Contactor 2	3	0.9	2.1	3.5	0.45	0.0
Main Ozone Contactor 3	3	0.9	2.1	3.5	0.45	0.0
Total Bromate Formed (μg/L)		15.7				0

This indicates that bromate formation is not a problem during average treatment plant conditions. If, however, the maximum ozone doses are applied during times when the water quality favours bromate production, then bromate formation is likely to be more severe

The modeling process can be used in reverse to plan a dosing regime for given water quality conditions that results in acceptable levels of bromate formation.

5 STATISTICAL ANALYSIS OF DATA VARIABILITY TO ASSESS PERCENTAGE COMPLIANCE TO A GIVEN LIMIT

So far bromate formation has been calculated for individual pairs of bromide and DOC data. The following statistical analysis enables an assessment of the likelihood of a given ozone dose resulting in a bromate level above a specified value using two input variables.

The following Monte-Carlo type procedure was adopted:

1. Collate available bromide and DOC water quality data for a given water treatment works ozonation process, and plot a histogram and identify the distribution that best fits the data (e.g. normal, or lognormal etc.)

2. Use the mean and standard deviation of the actual water quality data to generate model water quality distributions for bromide and DOC.

3. Set up two arrays, one bounded by the WQ model values, and the other bounded by the probability of the corresponding WQ having that value. In the body of the WQ array predicted bromate levels are calculated for the row and column WQ data points. In the body of the other array the probability of that WQ combination is calculated. The corresponding locations in each array give the probability of a given bromate concentration occurring in the ozonated water.

4. Tabulate the bromate formation and corresponding probability, and sort the data in ascending order according to bromate formed.

5. The cumulative sum of the individual probabilities up to a bromate level gives the total probability that the bromate formation does not exceed that level. An example graph of this output data is given below. In this case it shows that and ozone dose of 3 mg/L applied throughout the year will result in bromate levels below 15 μg/L for 90% of the time.

Figure 2   *Cumulative Probability of Given Levels of Bromate Formation*

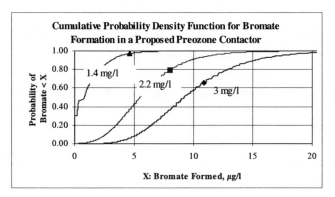

5.1 Limitations of this Analysis

The accuracy of the output depends on the accuracy of the water quality data model, and the bromate formation model. When this modeling was carried out, both these models were based on limited information. The subsequent Song *et. Al.* model cited earlier[6] is based on a data set of 640 experiments and so should provide improved accuracy.

6 IMPLICATIONS OF RISK OF BROMATE FORMATION ON OZONE CONTACTOR DESIGN

Before bromate was understood to be a problem, ozone contactor design aimed to achieve a residual of between 0.5 mg/L and 1.0 mg/L. A typical design had three chambers. In the first, sufficient ozone was dosed to satisfy the ozone demand of the water. In the second, ozone was dosed to maintain a residual of at least 0.4 mg/L in order to promote disinfection and/or pesticide oxidation. The third chamber was reserved for advanced oxidation. Ozone distribution nozzles are arranged to maximize ozone transfer to the water. Any local regions where the ozone concentration was higher than average were not perceived to be a problem.

The consensus now is that high ozone residuals should be avoided wherever possible, because bromate is formed whenever there is an ozone residual. A residual limit of 0.25 mg/L is used for design purposes.

Ozone contactor design is affected as follows:

1. Increased number of dosing points, with contact time between each point to allow residual to decay prior to next dose
2. Individual doses are reduced
3. Reduced nozzle density at dosing points to reduce local regions of relatively high ozone residual near the nozzles
4. Increased gas flow (i.e. reduced ozone concentration in dosing gas) to increase mixing in the bubble curtain, which again reduces local regions of high ozone residual
5. Apply ozone in sections of counter-current flow
6. Limit contactor tank depth to limit ozone transfer close to nozzles

The ozone contactors at a water treatment works were under construction as the bromate problem was becoming apparent. The design changes incorporated during construction are illustrated in figures 3 and 4 below.

Figure 3 *Typical Ozone Contactor Design Before Bromate Risk Was Identified*

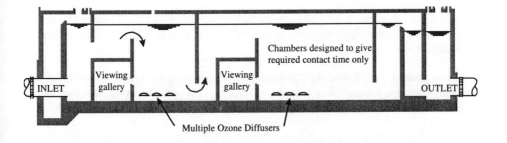

Figure 4   *Typical Ozone Contactor Design After Bromate Risk Was Identified*

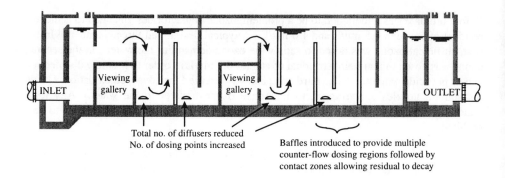

Total no. of diffusers reduced
No. of dosing points increased

Baffles introduced to provide multiple
counter-flow dosing regions followed by
contact zones allowing residual to decay

CONCLUSION

Bromate formation by ozonation at potable water treatment sites can be reduced to levels that are likely to be below the proposed legal limit of 10 μg/L through careful design, and dose control during operation. Models can be used both to check designs for possible bromate formation, and to set dosing regimes in contactors that have already been constructed to ensure that compliance limits are met.

Montgomery Watson acknowledge with thanks the assistance of Essex and Suffolk Water plc in the preparation of this paper.

References

1 Shukiary H.M., Miltner R.J. and Summers R.S.; "Bromide's effect on DBP formation, speciation, and control, Part 1: Ozonation", *JAWWA* **86**,No 6 pp 72-87,June 1994

2 Amy, Siddiqui, Ozekin and Westerhoff; "Threshold Levels For Bromate Formation In Drinking Water", *Water Supply*, **13**, No 1, Paris, pp 157-167, 1995.

3 Kruithof, J.C., Meijers, R.T., "Bromate formation by ozonation and advanced oxidation and potential options in drinking water treatment", *Water Supply,* **13** No 2, South Africa, pp. 93-103, 1995

4 Urs von Gunten, and Hoigné, J., "Bromate Formation during Ozonation of Bromide-Containing Waters: Interaction of Ozone and Hydroxyl Radical Reactions", *Environ. Sci. Technol.* **28**, 1234-1242, 1994

5 Siddiqui, M.S., and Amy, G.L. "Factors Affecting DBP Formation During Ozone-Bromide Reactions", *J. AWWA* **85,** p70, January 1993

6 Song, R., Minear, R., Westerhoff, P. and Amy, G., "Modeling and Risk Analysis of Bromate Formation from Ozonation of Bromide-Containing Waters", *Wat. Sci. Tech.,* No. 7-8, pp. 79-85, 1996.

PRODUCTION OF DRINKING WATER WITHOUT CHLORINE: THE DUTCH EXPERIENCE

J.P. van der Hoek[1], A. Graveland[1], J.A. Schellart[1], R.A.G. te Welscher[1], H.S. Vrouwenvelder[2] and D. van der Kooij[2]

[1]Amsterdam Water Supply, Vogelenzangseweg 21, 2114 BA Vogelenzang, the Netherlands; [2]Kiwa Research & Consultancy, Groningenhaven 7, PO Box 1072, 3430 BB Nieuwegein, The Netherlands

1 INTRODUCTION

Amsterdam Water Supply (AWS) is the oldest water supply company in the Netherlands. It started in 1853, using natural ground water from the dune area West of Amsterdam. At the moment, the total production capacity is 101 million m^3/year and surface water is used as raw water source. The drinking water is produced in two plants: production plant Weesperkarspel, which has a capacity of 31 million m^3/year, and production plant Leiduin, which has a capacity of 70 million m^3/year. Since 1983, no chlorine has been used in both plants, nor in the treatment process as final disinfection, nor in the distributed water as safety disinfection. The conditions how this can be realized will be described in this paper, focussing on the Leiduin plant.

2 THE LEIDUIN PLANT: PROCESS SCHEME AND CHARACTERISTICS

The process scheme of the Leiduin plant is shown in Figure 1. River Rhine water is extracted from the Lek Canal in Nieuwegein, in the centre of the Netherlands. There the water is pretreated by coagulation-sedimentation-filtration. Then the water is transported over a distance of 55 km to the dune area West of Amsterdam where artificial recharge is carried out. After a residence time of 100 days the water is abstracted and collected in an open reservoir, the "Oranjekom". Subsequently the water is treated by rapid sand filtration (for removal of suspended solids, algae, iron and ammonia), ozonation (for disinfection and to increase the biodegradation of organic compounds in the carbon filters), pellet softening (for hardness reduction), biological activated carbon filtration (for removal of organics, a.o. pesticides and AOX) and finally slow sand filtration (polishing filtration for final AOC-removal -easily assimilable organic carbon- and additional disinfection).

A main characteristic of this process scheme concerns the production of hygienically safe and biologically stable water without the use of chlorine. The hygienic quality of the water is in compliance with the regulations, and the nutrient level is low enough to prevent regrowth of bacteria in the distribution network. So, an important question is: how can this be realized? To explain this, the theoretical approach has to be defined, and the experiences in practice have to be described.

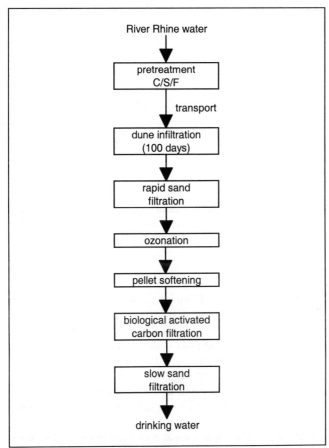

Figure 1 *Process scheme of the drinking water production plant Leiduin*

3 THE THEORETICAL APPROACH

In the theoretical approach, a distinction is made between the final disinfection and the safety disinfection. The final disinfection concerns the removal and inactivation of micro-organisms and is achieved by ozonation and slow sand filtration. Table 1 shows that with the available process units in theory all aspects are covered. The safety disinfection concerns the prevention of regrowth of micro-organisms during storage and distribution. This is achieved by removal of nutrients: N and P compounds in the pretreatment, C compounds in the biological activated carbon filters, and AOC in the biological activated carbon filters and in the slow sand filters. Table 2 shows that all aspects are covered with the available process units.

Table 1 *The final disinfection*

	bacteria	viruses	cysts	
			Giardia	Cryptosporidium
ozonation	+	+	+	±
slow sand filtration	+	±	+	+

Table 2 *The safety disinfection*

	N	P	C	AOC
pretreatment C/S/F	+	+	±	-
biological activated carbon filtration	±	±	+	±
slow sand filtration	±	±	±	+

4 THE PRACTICAL EXPERIENCE

The results of the Leiduin plant have shown that this theory can be applied in practice. This will be shown by describing the hygienic quality and the biological stability of the treated water, and of the water in the supply area after transport and distribution.

4.1 The treated water

4.1.1 Hygienic quality of the treated water. From long time experience it is known that the hygienic load of the treatment process is higher in winter than in summer, due to concentration of birds in the open collecting reservoir the "Oranjekom". The main process variable to increase the disinfection capacity of the plant is the ozone dose: in summer an ozone dose of 0.75 mg/l is applied and in winter an ozone dose of 0.9-1.0 mg/l, resulting in a CT-value of 0.5 mg.min/l and 2 mg.min/l respectively. Main questions are:

- is it possible to maintain the hygienic quality of the treated water during winter?
- does the elevated ozone dose result in too high bromate concentrations in the treated water?

Concentrations of pathogens in the water after ozonation as measured in the winter period 1995-1996 are summarized in Table 3. Already after ozonation the hygienic quality is in compliance with the legislation. The quality with respect to Giardia cysts and Cryptosporidium oocysts was estimated by measuring the concentrations in the raw water ("Oranjekom"), assuming an elimination capacity of the treatment process based on literature and particle counting measurements, and calculating the concentrations in the treated water based on the raw water concentrations and the assumed elimination capacity. The calculated concentrations in the treated water were compared with the maximum admissible concentrations based on a risk level of 10^{-4}. The calculations, summarized in Table 4, show that the concentrations in the treated water comply with the maximum admissible concentrations (MACs).

From previous pilot plant experiments[1] it is known that the raw water characteristics (DOC 2 mg/l, Br⁻ 150-200 µg/l, pH 8.0) restrict the ozone dose to approximately 1 mg/l to keep the bromate concentration below 5 µg/l, the proposed maximum admissible concentration in the Dutch legislation. Figure 2 shows the ozone dose, the CT-value and bromate concentrations in the treated water. At a high ozone dose the bromate concentrations can be restricted most times to 5 µg/l (an ozone dose of 0.25-0.5 mg/l was applied during the start-up of the new ozonation plant).

Table 3 *Effect of ozonation on the hygienic quality (CT = 2 mg.min/l, median values)*

	rapid sand filtrate	after ozonation	MAC
Coliforms 44 ^{0}C (n/100 ml)	13	<0.1	<0.3
fecal Streptococci (n/100 ml)	15	0.3	<1
Clostridia (n/100 ml)	1.0	0.2	<1

Table 4 *Effect of the treatment process on Giarda and Cryptosporidium*

	raw water[a] (n/l)	elimination capacity ($\log^{10}$ units)	treated water quality (n/l)	MAC (n/l)
Giardia	0.1	6	$0.1*10^{-6}$	$6.7*10^{-6}$
Cryptosporidium	0.24	4	$2.4*10^{-5}$	$3.3*10^{-5}$

[a]"Oranjekom" water

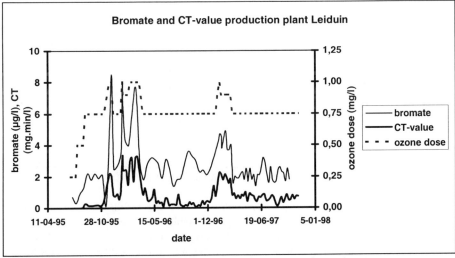

Figure 2 *Effect of the ozone dose on CT-values and bromate formation*

4.1.2 Biological stability of the treated water. To find out whether the treated water is stable, AOC concentrations and the Biofilm Formation Rate (BFR) were measured. AOC concerns the easily assimilable organic carbon. Water is assumed to be biologically stable when the AOC concentration is below 10 µg/l². Another good system to measure the biological stability of water is the biofilm monitor. In this system, which consists of a verticle glass column containing glass cylinders, a flow rate of 0.2 m/s is maintained. The BFR is measured by determining the concentration of ATP on the glass cylinders and is expressed in pg ATP/cm².day. In contrast with the AOC method which uses two types of selected bacteria strains, in this biofilm monitor system also nutrients like ammonium and methane contribute to the regrowth potential. It has been shown that water treated by slow sand filtration has a BFR lower than 1 pg ATP/cm².day. A BFR below 10 pg ATP/cm².day corresponds well with a low concentration of *Aeromonads* in the treated water[3,4]. An additional benefit of the biofilm monitor system is the possibility to measure the deposition of iron and manganese in the biofilm. These can be expressed in mg Fe/m^2.day and mg Mn/m^2.day.

Figure 3 shows the AOC concentrations as measured in the subsequent treatment steps of the Leiduin plant. After ozonation the AOC is increased to 60-140 µg/l. After having passed the two stage carbon filtration and the slow sand filtration the AOC is always below 10 µg/l. The biofilm monitor system, fed with treated water, showed that the BFR of the finished water is 0.25 pg ATP/cm^2.day while the deposition of Fe and Mn were below the detection limit. Both the AOC concentrations and the BFR imply biologically stable water.

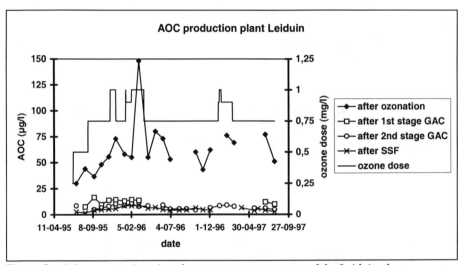

Figure 3 *AOC concentrations in subsequent treatment steps of the Leiduin plant*

4.1.3 Evaluation of the treated water quality. From the results it is clear that the treated water is hygienically safe and biologically stable. Although ozonation is used for final disinfection, the concentration of the by-product bromate in the finished water is always below 5 µg/l, also in winter when a relatively high ozone dose of 0.9-1.0 mg/l is applied to obtain a higher disinfection capacity. So, from the point of view of the treated water quality, the use of chlorine is not necessary with the process set-up as applied by AWS.

4.2 Changes during transport and distribution

Although the water is hygienically safe and biologically stable after treatment, changes may occur during transport and distribution. Therefore, AWS measured several parameters in the distribution network, a.o. AOC concentrations, heterotrophic plate counts (3d 22 °C), Aeromonads (30 °C) and biofilm characteristics.

4.2.1 Characteristics of the network. The characteristics of the network can affect the quality of the water during distribution. Hence, the characteristics of the network are summarized. The total length of the mains and distribution pipes is 1,892 km, and the internal surface area is 1,237,045 m². Cast iron (not protected inside) and PVC comprise most of the material (45% and 25% respectively). It is a looped network with an average residence time of 24 hours, and a maximum residence time of 2-3 days. The system is always kept under pressure and the leakage is less than 3%.

4.2.2 AOC concentrations in the distribution network. To examine whether AOC concentrations change during transport and distribution, AOC concentrations were measured in the supply area and were compared with the AOC concentrations in the treated water, directly after the transport pumps. Table 5 shows that no changes occur during transport and distribution.

Table 5 *AOC concentrations in the treated water and in the supply area*

	AOC in treated water (µg/l)	AOC in the supply area (µg/l)
April 1996	6	5
July 1996	4	5
September 1996	4	7
January 1997	7	5
February 1997	7	5
August 1997	3	4
Average	5.1 ± 1.7	5.2 ± 1.0

4.2.3 Heterotrophic plate counts (HPC) and Aeromonads. Figure 4 shows the HPC as measured in the treated water directly after the transport pumps (Leiduin 1 and Leiduin 2) and in the supply area. The very low numbers of HPC in the treated water show no increase in the distribution network. Although not shown here, also the number of *Aeromonads* are very stable during the whole year and do not increase in the distribution network.

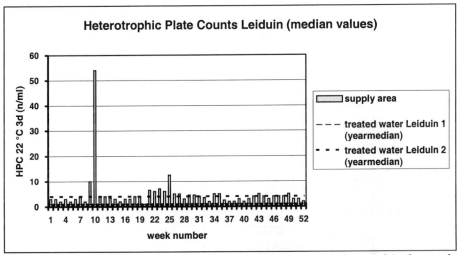

Figure 4 *Heterotrophic plate counts in the treated water at the plant and in the supply area*

4.2.4 Biofilm characteristics.   The biofilm characteristics as measured in the treated water at the production plant Leiduin were compared with biofilm characteristics in the supply area. For this reason pieces of PVC pipes were taken out of the distribution network at several locations, and the biofilm chracteristics on the inner surface were determined. It appeared that in the distribution network the concentration of biomass and iron were significantly increased: in the treated water the biofilm contained 40 pg ATP/cm^2 and 3.4 mg Fe/m^2, while the maximum values in the network were 1020 pg ATP/cm^2 and 700 mg Fe/m^2. The increase correlated with the length of cast iron pipes the water had flown through. Taking into account the relatively high chloride concentration (100-130 mg/l) we think that corrosion effects may contribute to the development of a biofilm in the network. Further research is being started up at the moment at AWS.

4.2.5 Evaluation of the changes in water quality during distribution. The results show that during distribution the hygienic quality and biological stability of the water can be maintained. Points of concern are the increased ATP and iron concentrations of the biofilm in the distribution network, although this does not affect the quality of the water. Also from the point of view of distribution it can be concluded that the use of chlorine can be omitted in a process set-up and distribution system as applied by AWS.

5 NEW DEVELOPMENTS

At the moment several water supply companies in the Netherlands examine the feasibility of using membrane filtration in the treatment schemes[5]. These schemes are refered to as Integrated Membrane Systems (IMSs): membrane filtration processes as part of the whole treatment scheme and integrated in the whole process.

Examples of these IMSs are the use of reverse osmosis by AWS[6,7,8] and the use of ultrafiltration and reverse osmosis by the PWN Water Supply Company North Holland[9]. This second example is already being realized at full-scale (18 million m^3/year) and will be in operation from 1999. Figure 5 shows the IMS of PWN. After conventional pretreatment, water from the IJssel-lake is treated by a dual-membrane process consisting of ultrafiltration and reverse osmosis.The treated water is mixed with water from conventional treatment plants.

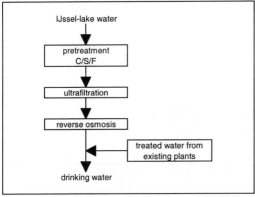

Figure 5 *Integrated Membrane System of PWN*

The use of reverse osmosis at AWS is shown in Figure 6. It is used in an extension scenario for the Leiduin plant. In the so called "direct treatment scheme", 13 million m^3 per year of pretreated River Rhine water (C/S/F) will be treated without soil passage in the dune area, but directly by a scheme using ozonation, biological activated carbon filtration and slow sand filtration as pretreatment for the final step reverse osmosis.

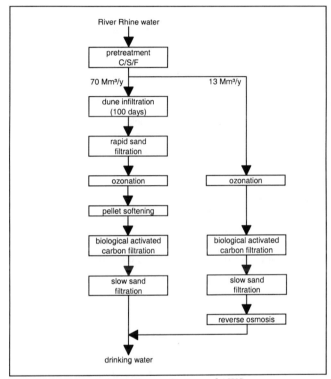

Figure 6 *Integrated Membrane System of AWS*

Research has shown that these IMSs can contribute to the production of hygienically safe and biologically stable water. With respect to micro-organisms, a reduction capacity can be calculated of 9 log^{10} units. Particle countings, challenge tests using MS2-phages and literature[10] have confirmed this high capacity. AWS measured the biological stability of the permeate. An extremely low AOC concentration of 0.5 µg/l was measured, while the BFR was also extremely low, 0.006 pg ATP/cm^2.day. So, IMSs result in additional possibilities to produce and distribute water without the use of chlorine.

6 CONCLUSIONS

The presented data show that it is possible to produce hygienically safe and biologically stable water without the use of chlorine. Prerequisites are that the elimination capacity of the treatment process is high enough to inactivate all pathogenic micro-organisms, and that the nutrient level in the treated water is low enough to prevent regrowth of micro-organisms. The distribution network must always be kept under pressure to avoid contamination from outside, while the materials used must not contribute to regrowth of micro-organisms. Introduction of Integrated Membrane Systems in the drinking water industry enlarges the possibilities to produce drinking water without the use of chlorine.

References

1. J.P. van der Hoek, E. Orlandini, A. Graveland and J.G.M.M. Smeenk. *Proceedings IWSA Congress and Exhibition*, 1995, **20**, SS5-1.
2. D. van der Kooij. *Journal AWWA*, 1992, **84** (No. 2), 57.
3. D. van der Kooij, H.S. Vrouwenvelder and H.R. Veenendaal. *Water Science & Technology*, 1996, **32**, 61.
4. D. van der Kooij, H.S. Vrouwenvelder and H.R. Veenendaal. *H₂O*, 1997, **30**, 767.
5. J.C. Kruithof, P. Hiemstra, P.C. Kamp, J.P. van der Hoek, J.S. Taylor and J.C. Schippers. *Proceedings IWSA Congress and Exhibition*, 1997, **21**, SS1-16.
6. J.P. van der Hoek, P.A.C. Bonné, E.A.M. van Soest and A. Graveland. *Proceedings AWWA Membrane Technology Conference*, 1995, 277.
7. J.P. van der Hoek, P.A.C. Bonné, E.A.M. van Soest and A. Graveland. *Proceedings AWWA Membrane Technology Conference*, 1997, 1029.
8. J.P. van der Hoek, P.A.C. Bonné, E.A.M. van Soest and A. Graveland. *Proceedings IWSA Congress and Exhibition*, 1997, **21**, SS1-11.
9. P.C. Kamp. *Proceedings AWWA Membrane Technology Conference*, 1995, 31.
10. J.C. Lozier, G. Jones and W. Bellamy. *Journal AWWA*, 1997, **89** (No. 10), 50.

Concluding Remarks

DISINFECTION BYPRODUCTS - THE WAY FORWARD
SUMMARY OF CONFERENCE

Dr R A Breach
Head of Quality & Environmental Services, Severn Trent Water
and Chairman, EUREAU Commission 1

The Conference was an undoubted success, and provided a very useful overview of the current position with disinfection byproducts (DBPs) in Drinking Water. The proceedings have benefited greatly from the wide range of experience of the different delegates, from both the USA and Europe. It was particularly useful to have delegates from utilities, regulatory agencies, research organisations and equipment suppliers involved in constructive debate. One speaker used the phrase "one man's fact is another man's opinion". This Conference has sought to distil out those aspects which are agreed facts, and separately identify those issues where judgement has to be exercised.

A number of key themes emerged from the discussion, particularly the need to balance increasingly complex and potentially conflicting objectives in water treatment and distribution. To achieve an overall good water quality for customers, utilities have to be aware of all potential interactions and constraints, and recognise that disinfection byproducts are only one part of good quality management.

Good objective research provides vitally important input to the DBP debate. Equally the results of such research must be kept in proportion. The need for a reality check is always useful when debating the significance of research in this field. There is no doubt that more and more research on lower and lower levels of disinfection byproducts can potentially provide more to worry about, but equally it allows progressive refinement of practicable DBP reduction strategies.

It is always of critical importance to look at issues from a customer viewpoint, rather than solely that of a technocrat. Customers' priorities are safe, pleasant and affordable water. Their highest priority would always go to water free of pathogenic organisms, but also ideally free of disinfectant tastes. Disinfection byproduct levels usually have the lowest priority, mainly because the public has insufficient technical knowledge to make an informed judgement. With surface water sources the costs of achieving strict quality standards can be high and perhaps very high, depending on how low a level of disinfection byproducts it is necessary to achieve. On this basis it is interesting to debate whether water which achieved high microbiological quality, and was free of disinfectant tastes, would by default have achieved sufficiently low levels of DBPs to protect public health.

What has emerged from this Conference is a reminder of the well established principle that achieving good water quality is more than just good treatment. We continually have to remember that cost effective management of water quality involves all aspects, from raw water catchment through to the customer's tap, including active control of the distribution network.

We have also learned much about the way in which different treatment processes can be used to achieve low disinfection byproducts, as well as all the other water quality objectives. Experience over many years has shown that there is no treatment "magic bullet". All treatments have both strengths and weaknesses, and before introducing any new process there has to be clarity about the objectives, and also a good understanding of the potential constraints. Any new treatment techniques must be introduced progressively, so that any potential problems, as well as benefits, can be properly managed. It is certainly becoming evident that the expectations of both customers and regulators are demanding ever more reliable treatment. It is not the 99.99% of the time that treatment operates perfectly, that is of interest to them; it is the 0.01% of the time that things might go wrong. To achieve very high levels of treatment reliability, there is a growing awareness that better use should be made in the future of quantitative risk management tools.

The Conference provided the expected discussion about whether chemical disinfection has a future. Certainly we heard that there are some parts of Europe where high quality microbiological water can be achieved, with zero or minimal use of chemical disinfectants. However, achieving this goal, at least with surface water, demands a high level of investment and very active management of the whole water quality process. The Conference reinforced the message that disinfection must always be paramount. The feeling was that whilst, in the long term, it may be possible to more widely phase out or further reduce the use of chemical disinfectants, at least in the more developed countries, this would not happen overnight, and in many systems chemical disinfection could be with us well into the foreseeable future.

In conclusion, there were perhaps five key areas of agreement:

1   We must maintain co-operation between all the different parties involved. Consensus regulation works. That means that through proper dialogue between regulators, utilities, customers, researchers and other stakeholders, the optimum balance can be struck between high levels of public protection, and practicable but affordable quality standards.

2   The availability of good research allows better regulatory decisions, but research should be targeted on those issues that allow better decision making.

3   All practical experience suggests that water treatment is an evolutionary, not revolutionary, process. We need to optimise our existing treatment regimes, and

progressively introduce any new processes based on a thorough understanding of the advantages and disadvantages that this will bring.

4   We should never forget that water quality management is a holistic process. We have many potentially conflicting quality objectives, and the job of the water quality manager is to balance all of these to achieve a water which is of all-round acceptable quality.

5   Finally, we must never forget our customers. They are not experts in disinfection byproducts, but they do demand safe, pleasant and affordable water. Through the excellent co-operation demonstrated in this Conference, I believe we are well placed to move forward to meet this goal.

Subject Index